John Ellor Taylor

Geological Stories

A series of autobiographies in chronological order

John Ellor Taylor

Geological Stories

A series of autobiographies in chronological order

ISBN/EAN: 9783337012984

Printed in Europe, USA, Canada, Australia, Japan

Cover: Foto ©berggeist007 / pixelio.de

More available books at **www.hansebooks.com**

GEOLOGICAL STORIES.

GEOLOGICAL STORIES:

A

SERIES OF AUTOBIOGRAPHIES

IN

CHRONOLOGICAL ORDER.

BY

J. E. TAYLOR, F.G.S.,

AUTHOR OF "HALF-HOURS AT THE SEASIDE," ETC.

LONDON:
ROBERT HARDWICKE, 192, PICCADILLY, W.
1873.

PREFACE.

The "Stories" whose collection make up this little volume originally appeared in 'Science Gossip' and other magazines. They are now arranged in their geological order, so that it has been attempted to give as simple and picturesque a view of the past history of the globe as possible. The favourable reception given to the "Stories" during their separate publication was an encouragement to their issue in the present form. The illustrations which now appear are an addition, and, it is hoped, they will assist in rendering that part of the text which may be confused more intelligible.

The author feels sure that scientific men are too anxious to have their audiences increased, to quarrel with any honest means by which that desired end may be brought about. In the publication of these "Stories," no other object has been in view than that of trying to make the number of people who were prejudiced against, or careless about, the grand deductions of Geology interested in a science which is now influencing almost every department of modern thought.

CONTENTS.

CHAPTER I.

THE STORY OF A PIECE OF GRANITE.

PAGE

Discussions on the origin of granite—Plutonists and Neptunists—Victory of the former—Result of carrying the victory too far—Mineralogical composition of granite—*Quartz*, its component parts—Alkaline ingredients probably assisted in its fluxion—*Felspar*, what it is composed of—Its "rottenness"—Origin of *Kaolin*, or "China-clay"—*Mica*, description of—*Hornblende*, its colours, chemical composition, &c.—Mineralogical relations of granite—*Porphyry*, origin of its name—Relation of ditto to other igneous rocks—*Syenite*—*Protogine*—*Pegmatite*—Binary granites—Ternary ditto—Quaternary ditto—Their classification—The first formed crust of the globe—Idea as to the origin of the metamorphic rocks—Conditions under which granites were originally formed—Cavities in the quartz crystals of granite—Heat and water concerned in origin of granite—Pressure under which granite was formed—Consequences of absence of pressure on igneous rocks—The granites of the Highlands—Pressure under which they were formed—Ditto of the granites of Cornwall—Difference of rock-pressure—Great antiquity of the earth proved by above facts—Granite probably formed of re-melted stratified rocks—Deep-seated condition of latter—Upheaval of granite nucleus of mountains—How it is granite now occupies large areas—The denudation of rocks formerly overlying granite outcrops—Internal force of the earth as opposed to that of the sun—Overlying rocks stripped off by atmospheric agencies—Time occupied . . . . 1—13

CONTENTS.

CHAPTER II.

THE STORY OF A PIECE OF QUARTZ.

Time as a factor in Geology—Astronomical calculations—Our notions of the earth's antiquity greatly expanded within last ten years—The *Cambrian* period—"Primary" rocks—Possibly older stratified rocks than any yet known—Quartz rock, its external appearance—Crystals of ditto—*Quartz* and *Quartzite*—Difference of ditto—"Brazilian pebbles"—The *Laurentian* period the oldest yet known—Composition of *Quartz*—How difference in colour was effected—*Amethyst*—*Topaz*—*Cairngorms*—*Jasper*—*Flint*—*Chalcedony*—*Agates*—The metamorphic rocks—Former notion of ditto—Absence of fossils in ditto—Notion as to the Cambrian period—Altered rocks—Cause of ditto—Poverty of species of organic forms in older rocks—What quartz originally was—How formed—The *Eozoon*, or "Dawn-animalcule"—Organization of ditto—Its natural history habits—Recollections of a Piece of Quartz—*Plumbago*, or "black-lead"—How formed—Lowly organization of earlier land plants—The Laurentian limestones—How they were probably formed—Phosphates of lime—Thickness of lower Laurentian rocks—Changes during Laurentian period—Elevation of marine deposits into dry land—A period of submergence—The upper Laurentian rocks—How formed—"Unconformability" illustrated—Inferences from such phenomena—Changes through which Laurentian rocks have passed—Denudations of ditto—Intrusions of igneous rocks in ditto—Metamorphism of sandstones, limestones, &c.—Contortion of rocks . . . . 14—23

CHAPTER III.

THE STORY OF A PIECE OF SLATE.

Locality of slate rocks—Scenery of the Welsh mountains—Original condition of latter—Difference in mode of formation of sandstone and limestone rocks—"Cleavage," what it is—Probable cause of ditto—"Joints" in slate rocks—Contortions in *Cambrian* rocks—Wear and tear of ditto—Origin of the contortions, &c.—Experiments on rock contortion—The *Protozoic* period, what it was thought to be—Age of Cambrian rocks—Fossils of ditto—Laws of the succession of life-forms—Lowly organization of Cambrian fauna and flora—Thickness of Cambrian rocks—Composition of ditto—Evidences of shallow sea deposition—How such a thickness of rock material could thus be formed—*Arenoolites*, or fossil worm-borings—Ripple-marks in Cambrian strata—Rain-pittings and sun-cracks in ditto—The first crustaceans—Zoophytes of the Cambrian seas—The Lower Cambrian formation as a group of rocks—Break in continuity of Cambrian rocks—Absence of a great limestone deposit—Where latter will possibly be found—The upper Cambrian rocks—*Lingula* flags—The brachiopodous mollusca—Antiquity of certain forms of life—Fossil shrimps—Ancient sea-lilies—Habits of ditto—The *Trilobites* — Pteropods—Why Cambrian fossils are found as "casts"—Influx of higher forms of life—The *Cephalopodous* mollusca— *Orthoceratites* — *Gasteropoda*, or marine snails—The *Bellerophon*—Paucity in specific types—Progression of life system—Inference from ditto . . 24—36

CHAPTER IV.

THE STORY OF A PIECE OF LIMESTONE.

Limestones common to every geological formation—Generally of *vital* origin—Evidences of deep sea origin—Characterized by abundance of marine organisms—Sandstones contain more terrestrial ditto—Silurian limestones in the "Black Country"—The Wren's nest. Dudley—Abundance of fossils—Silurian limestones in America—Origin of *Petroleum*, or "rock-oil"—The Wenlock limestones—*Trilobites*, abundance of species of—Alternations in stratification of Silurian rocks—Shingle-beds—Thickness of Silurian rocks—Greater distribution of sea areas—Probable height of mountains—The Lower Silurian rocks—Evidences of volcanic disturbances—Bala limestones—Ca·adoc sandstones—The Middle Silurian rocks — May Hill sandstones — Tarannon shales—Upper Silurian rocks—Their thickness, &c. — Woolhope beds — Wenlock limestones and shale — Ludlow beds — Aymestry limestones—Downton sandstones—Marine animals of the Silurian period—*Orthoceratites*—"Sea-pens"—Graptolites—Structure of the *Trilobites*—Inferences of ditto

CONTENTS.

—Great change in physical geography during Middle Silurian period—Thickness of volcanic lava and ash beds—Abundance of marine life during Upper Silurian times—*Pectens* — Univalves — *Cystideans* — The "chain-coral"—Ancient coral reefs in Shropshire, &c.—*Favosites polymorpha*—Brachiopodous mollusca of Silurians—First appearance of Vertebrate animals as fishes—Ancient feeding-grounds of ditto—The Ludlow "Bone-bed"—Recent bone-bed off the Irish coasts—Presence of fossil spores of cryptogamous plants in Ludlow bone-bed—Inferences from ditto—Origin of metal lodes—Usual preponderance of gold in Silurian rocks, &c. 37—57

CHAPTER V.

THE STORY OF A PIECE OF SANDSTONE.

Sandstone beds in every geological formation—*Psychometry*—How sandstone rocks were originally formed—The physical changes that have taken place since—Principal agents in formation of sandstones—Origin of Old Red Sandstone—Cementing agents in sandstone rocks—Free-stones and flag-stones—The *Devonian* period—Rocks of ditto in America, South Africa, Russia, Asia-minor, and Australia—Origin of the name *Old* Red Sandstone—Thickness of ditto—Divisions of ditto—Varying thickness in different localities—Error as to greater prevalence of Carbonic acid in atmosphere during *Palæozoic* epoch—How the rocks were coloured red—Ancient sea-weeds—Huge Crustaceans—*Pterygotus*—"Age of fishes"—Structure of first order that appeared, the *Ganoids*—Relation of ditto to existing species in America and elsewhere—The *Asterolepis* — *Holoptychius*— *Pterichthys*—*Cephalaspis*— *Coccosteus*, &c.—External peculiarities of ganoid fishes—The *Placoid* fishes—*Onchus*—Fossil fishes of Caithness, &c.—Shoals of fossil fish—How accumulated—Reptilian characters of some Devonian fishes—Corals, Trilobites, &c.—Differences in physical condition of old sea-bottom—Chief localities of Devonian fossiliferous strata—Total number of species—The "Piltcn group"—*Brachiopodous* shells—*Spirifer*—*Clymenia*—"Sun-corals," &c.—Coloured fossil corals of Torquay, &c.—Peculiar Trilobites of Devonian rocks—Devonian rocks in Ireland—Evidences of fresh-water deposition—The "Irish primitive Fern" (*Palæopteris Hibernicus*)—Its structure—Other Devonian plants—*Sagenaria*—*Psilophyton*—Abundance of fossil fresh-water mussels (*Anodonta*)—Remains of crustacea—Resemblances of Devonian lakes to existing North American ditto . . . . 58—78

CHAPTER VI.

THE STORY OF A PIECE OF COAL.

The piece of coal commences its history—How many periods elapsed since it was formed—Their names—Physical conditions of Carboniferous period—Reptiles of ditto—Ancient coral reefs—Formation of "Mountain" or Carboniferous Limestone—Its abundance of fossils—The principal species —*Nautilus*—*Goniatites*—*Gyroceras*—*Orthoceras*, size and habits of—Abundance of ditto—Shoals of *Spirifera*—Their structure—Carboniferous Trilobites—Their last appearance —The Crinoids—Derbyshire *encrinital* limestones—Polyzoa of Carboniferous period—Fishes of ditto—*Megalichthys*—*Palæoniscus*, &c.—Forests of later Carboniferous period—Their appearance—Principal vegetable forms of ditto—*Lepidodendra*—*Sigillariæ*, &c.—Gigantic Club-mosses—*Calamites*, or gigantic "Horse-tails"—Coniferous trees—Fossil ferns—Accumulation of vegetable materials—Chemical changes in ditto—Gradual change to present state as coal—Microscopical revelations of coal-ash—Changes through which carbon has passed—Gradual transition from wood to anthracite—Correlation of physical forces—Light and heat of the Carboniferous period "bottled-up" in coal—Colours of original plants, &c.—Economical advantages of coal to mankind . . . . 79—113

CHAPTER VII.

THE STORY OF A PIECE OF ROCK-SALT.

Difference of story-teller from its fellows—Natural origin of rock-salt—Interval between Carboniferous and Triassic period occupied by the *Permian*—Characters of latter—Evidence of ancient Ice-action—Permian "breccias"—Re-

semblance of its fossils to Carboniferous forms—Distribution
of Rock-salt—The *New Red Sandstone*, or "Trias"—The
brine-springs of Cheshire—Divisions or Trias—Bunter sandstone—*Muschelkalk*, or "shelly limestone" of Germany—
Keuper beds—A tolerably deep sea in Germany—"Lily"
Encrinite—Reptilian fishes—"Breaks" in geological continuity—Hallstadt and St. Cassian beds—"Unconformability" of members of the Trias—Keuper beds those in
which rock-salt found — Cheshire strata — Rock-salt and
gypsum—Thickness of strata—Area of ditto—Evidence of
fish and reptile life—No fossils in salt-beds—An ancient
"Dead Sea"—Modern ditto—The *Keuper* seas — Pseu-
domorphic crystals of salt—Evidence of atmospherical and
mechanical action—Absence of vital ditto—A "Dead Sea"
in Cheshire and Worcestershire—Origin of the Cheshire
"meres"—How Rock-salt was formed—Dry-land appearance
of Triassic age—Great frog-like reptiles—*Labyrinthodon*—
Rhynchosaurus — Feet-impressions on sandstone — Extinct
birds in America—First appearance of mammalia—*Microlestes*—Its existing relations—Rhætic beds—Ornithic affinities of reptiles—South African Trias—Its fossil reptiles—
Dicynodonts—Flora of Triassic period—Extinction of Palæozoic forms . . . . 114—127

CHAPTER VIII.

WHAT THE PIECE OF JET HAD TO SAY.

Where jet occurs—*Lias* beds—Origin of jet—Ditto of name
"Lias" — Stratigraphical appearance of Liassic sea—The
"struggle for life"—Thickness of the Lias beds—Division
of ditto—Lower Lias shales and limestones—Dry land of the
Liassic age—Liassic flora—Introduction of new forms—The
"Age of Reptiles"—*Ichthyosaurus* — *Plesiosaurus* — Great
land reptiles — Flying Lizards — *Pterodactyles* — Physical
geography of Lias sea—Habits of marine lizards—Fish of
Liassic era—*Lepidotus*—*Dapedius*—*Æchmodus*—*Hybodous*
—Ancient sea-lilies—*Pentacrinus*—Description of ditto—
Sea-bed of Lias period—Abundance of Cephalopods—*Ammonites*—*Nautilus*—*Belemnites*, or "thunder-bolts"—Nume-
rical abundance of Ammonites—Ditto of Belemnites—Census

CONTENTS. xiii

of Liassic fossils—Brachiopods—Last appearance of *Spirifera*—Numerical proportions of Conchifera—*Gryphæa*—*Hippodium*—*Avicula*, &c.—Plants and insects of the Lias—Continuity of the great Life-scheme—The Plan of Creation . . . . 128—145

CHAPTER IX.
WHAT A PIECE OF PURBECK MARBLE HAD TO SAY.

Use made of Purbeck marble—Its ecclesiastical importance—What formation it belongs to—The Oolite period—Subdivisions of Oolite rocks—Extinction and continuity of specific forms—Description of members of Oolite series—Physical geography resulting from their distribution—Geographical changes during Oolitic era—Oolitic coal shales near Scarborough—Physical conditions of the ancient sea-beds—Destruction of ancient encrinites—Dry land areas—Ancient lakes—Fossil fresh-water shells—*Planorbis*—*Paludina*, &c.—Accumulation of shelly limestones—Purbeck marble, how originally formed—"Coral Rag"—Ripple-marked flagstones—Evidences of shallow water—The great Oolite—Thickness of entire series—Fossils of the Oolite—Their variety—Description of Oolite seas—Marine reptiles—Ancient coral reefs—Ancient ganoid fishes—Mud-banks of Oolitic age—The flying reptiles—Terrestrial mammalia—Their variety—Flora of the Oolite—Oolitic coal, how formed—Oolitic iron-stone, how formed—The Stonesfield slates—Number of species of mammalia—Marsupial animals—Solenhofen limestones—Their richness in organic remains—Cycads and Zamias—Portland "Dirt-bed"—How formed—" Birds' nests "—Great land reptiles—*Megalosaurus*—" Missing links "—Bird-like affinities of Oolitic reptiles—How many of them were two-legged—*Compsognathus*—Fossil reptilian eggs—Gradual development of Reptilia—The first Bird—*Archæopteryx*—Its reptilian affinities—Physical changes since the Oolitic period—The Himalayahs—All changes tending towards a higher condition of existence . . . . 146—166

CONTENTS.

CHAPTER X.

THE STORY OF A PIECE OF CHALK.

PAGE

The minerals diffused in sea-water—Marine animalcules—
The composition of their shells—Enormous power of multiplication—Chalk composed mainly of microscopic shells—
The *Globigerinæ*—Their antiquity—How chalk was formed
— *Coccoliths*—Atlantic mud — Changes along Cretaceous
sea-bed—*Silica*—How flint was formed—Flint-bands and
nodules—"Pot-stones" or *Paramoudræ*—Cretaceous Echino-
dernus — *Ananchytes*, or "Fairy loaves" — *Micrasters* —
Cidarids—"Thunder-bolts," or *Belemnites*—What they were
—Cretaceous Brachiopods—Cretaceous fish—Appearance of
new orders—A huge marine reptile—*Mososaurus* or *Leidodon*
—The *Wealden* series, how formed—Its thickness—Fossil
remains of ditto—Great land lizards—Subdivisions of chalk
strata—Characteristic fossils of ditto—Beauty of certain
fossils—How they have been preserved—Evidences of upheaval of ancient sea-bottom—Slowness of the process—The
forces of nature—Ancient sea-birds—How the newly-raised
land was peopled—Geographical changes subsequent to
deposition of chalk . . . . 167—192

CHAPTER XI.

THE STORY OF A LUMP OF CLAY.

Commonness of clay—How old clay strata have been altered—
The *Eocene* period—Difference between the London clay
and the Boulder clays—Circumstances under which the
London clay was formed—A tropical climate in Britain—
The commencement of the Tertiary epoch—Thickness of the
Eocene strata—Description of ditto—Their fossils—Geo-
graphical distribution of animals and plants—The flora of
the Eocene period—Fossil fruits—Indian scenery in Britain
—Mammalia of the Eocene period—*Palæotheria*—Ganoid
fishes—Boa-constrictors in England—Water-snakes—*Ano-
plotheria*—*Chæropotamus*, or "river-hog" — *Dichobune*—
Hyænodon—"Missing links"—Mollusca of Eocene period—
Their sub-tropical character—Sharks of the period—Turtles

—Duration of the Eocene period—Elevation of mountain chains—Physical geography of the Eocene period—Decrease in climature—The register of the earth's crust . . . 193—204

CHAPTER XII.

THE STORY OF A PIECE OF LIGNITE.

"Brown Coal," or *Lignite*, what it is—Its appearance, &c — The *Miocene* period—Great fresh-water lakes in Europe— Luxuriant flora of the Miocene period—Extent of ditto—No ice-cap at the North Pole—Connection of the Old and New Worlds—Number of flower-bearing plants found fossilized— Temperature of Miocene period—Its probable cause—Ancestry of living animals and plants—Lignite beds in Europe, Asia, &c.—Preservation of vegetable remains—Cosmopolitan character of the Miocene flora—The great number of *evergreen* plants—Distribution of Miocene species—The *American* character of fossil flora—*Smilax*—*Dryandroides*—*Proteacea*— Fan-palms — Tulip-trees — Magnolias, &c.—Beauty of the Miocene landscapes — Lignite beds of Bovey Tracey — Number of species of fossil plants—denudation of the Dartmoor granite—Miocene strata in Ireland and Scotland—The last active volcanoes in the British Isles—Lignite beds of Greenland, and what they teach—Ditto of Iceland—Ancient Miocene land—Great fresh-water lakes in Switzerland — Miocene strata of ditto—Their fossils—Fish, &c.—Caddisworms—Indian butterflies—*Termites*—Various fossil insects —Appearance of *Quadrumana* or monkeys in European woods—The *Dryopithecus*—*Semnopithecus* —*Pliopithecus*— Opossums in Europe —*Dinotheria* — Tapirs— *Mastodon* — Wart Hogs—Hippopotami—Wild Oxen—Wild Deer—The *Hipparion*, or three-toed horse—Miocene deposits in India —" Missing links" again—The *Sivatherium*—The great Indian tortoise — Giraffe and camel in India — Miocene horses in America—Mastodons and elephants in ditto—The long-armed monkey—Volcanoes in Central France—Upheaval of sea-beds—Formation of the Alps, Pyrenees, &c.—Refrigeration of climature—Formation of new sea-basins . . . . 207—221

CHAPTER XIII.

THE STORY OF THE "CRAGS."

Meaning of the term "crag"—Where these beds are found—Their number—The *Pliocene* period—The Coralline, Red, and Norwich Crags—Suffolk the bed of a tolerably deep sea—Fauna of the Coralline Crag sea—Ancient sea-urchins—Where now living—Time since Coralline Crag was formed—Abundance of mollusca in Coralline Crag sea—The *Astartes*—*Pectunculus*—*Cyprina*—Number of species of mollusca—An ancient sea-bed, and what it teaches—No "dredging" required—"Sea Mats"—*Fascicularia*—Extremes in Coralline Crag sea—Marine currents—The oldest crag—Physical conditions of Pliocene period—Migration of mollusca southerly—Coralline Crag shells in West Indies—Ditto in Mediterranean—Thickness of Coralline Crag—How formed—The *Red* Crag—Origin of name—Denudations of Coralline Crag previous to it—An ancient land surface in Suffolk and Norfolk—How the "coprolite" beds were formed—Mollusca of the Red Crag—"Southern" forms—"Northern" ditto—Refrigeration of climature—Corals—Cowrie shells—False current bedding in Red Crag—Physical geography of ditto—The chalk of Norfolk—A shallow estuary over site of Norwich—Evidence of river communication—The *Norwich* Crag—Its fluvio-marine character—Difference in fossils from previous crags—Abundance of littoral shells—*Tellina*—*Mactra*—*Cerithium*—Fresh-water shells—*Planorbis*, *Paludinæ*—*Lymnea*, &c.—Land-snails—Census of Norwich crag—Ancient land-animals in Norfolk—The Aldeby Crag—Its fossils—Increase of "Northerly" mollusca—Evidence of rigorous climate—The Upper Norwich Crag, and its lessons—*Pliocene* and *Pleistocene* periods—Analyzation of shells of crags—Number of extinct species in ditto—The Be'gium crags—The "Box-stones" of Suffolk—Their origin—Evidence of a broken-up deposit—Pliocene strata in Sicily—English shells in ditto—How they came there—Evidence of volcanic disturbance—Formation of Mount Etna—Height of Pliocene beds on its flanks—Refrigeration of climate in northern hemisphere—Preparation for the long Arctic winter—The introduction of the Glacial epoch   222—237

CONTENTS. xvii

CHAPTER XIV.
THE STORY OF A BOULDER.

PAGE

Its restless life—The Glacial period—Extent of Glacial deposits—"Noah's flood"—The "Northern Drift"—Connection of Tertiary life-forms with existing species—Sand, gravel, and clay—Evidence of extreme cold—Recapitulation of slow refrigeration of climate during Tertiary epoch—The Norfolk "Forest-bed"—Its fauna and flora—The bed of the German Ocean once a green wood—Strange animals which lived in ditto—An Arctic climate introduced into Britain—England under a wintry sea—Scotland and the north of England under an ice-sheet—Greenlandic circumstances in Britain—Description of Greenland glaciers—Icebergs—The Cromer cliffs—Thickness of Lower Boulder Clay—Iceberg action in drift—The Atlantic sea-floor—The Gulf stream—Depth to which England was submerged—How ascertained—Moel Tryfaen—An old sea-beach on ditto—Re-emergence of the land—Arctic mollusca in British seas—Raised sea-beaches—Stranding of ancient Icebergs—Dropping of large boulders—Migration of Arctic plants—The "chalky" Boulder Clay—The "Heavy Lands," how formed—Denudation of mud sheet into valleys—Post-glacial deposits—Ice-grooves and scratches—England still connected with the Continent—No straits of Dover yet formed—The *Mammoth*—The Hairy Rhinoceros—Land connection with Ireland and England—The Irish Elk—The Reindeer, Muskdeer—Lemming, &c.—Arctic plants living on British mountains—How they got there, and when—Glaciers in Scotland, Cumberland, Lancashire and Wales—The Swiss glaciers—The desert of Sahara once a sea—English mollusca inhabiting it—Formation of the desert sands—A warmer climate sets in—Migration of existing fauna and flora—Where from—Separation of Ireland from England—Of England from the Continent—Bone caves—Appearance of MAN—Flint implements—Migration and extinction of *Mammoth*, &c.—Formation of rich subsoils—Results of the glacial period . . . . 238—253

CHAPTER XV.

THE STORY OF A GRAVEL-PIT.

PAGE

The last of the race—Incompleteness of the story—Gravel-pits, and where—Valley or river gravels—How formed—River terraces—Difference in age of gravels—How detected —Peculiar appearance of pebbles in valley gravels—Heights of former river levels—Evidences of rigorous climate— Valleys scooped out—Evidences of man's first appearance— Flint implements in river or valley gravels—Proofs of their human workmanship—The antiquity of man—Identity of pattern in flint implements—*Palæolithic* types of ditto—contemporaries of primitive man—The *Mammoth*—Woolly-haired Rhinoceros, &c.—Why human bones not found with flint implements—Roman and Saxon cemeteries—Percolation of running water—Teeth and tusks most enduring—Stalagmite in limestones—How it preserves fossils—Human bones under ditto—Kent's Cavern—Evidence of human habitation of ditto—The "Reindeer" period—Bone caves in Southern France—Artistic attempts—The *Neolithic* period — How distinguished — Stone weapons — Distribution of Neolithic and Palæolithic implements—The Lake Dwellings —Retrospect—Evidence of Progression—Development of culture—The Higher Life . . . . 254—272

SUMMARY OF PREVIOUS CHAPTERS 273

APPENDIX, giving Table of Rock Strata in British Islands—Explanations and illustrations of geological terms, &c.—Conclusion 284
INDEX 293

GEOLOGICAL STORIES.

CHAPTER I.

THE STORY OF A PIECE OF GRANITE.

"We turned, we wound
About the cliffs, the copses, out and in,
Hammering and clinking, chattering stony names
Of shale and hornblende, rag, and trap, and tuff,
Amygdaloid and trachyte, till the sun
Grew broader towards his death, and fell, and all
The rosy heights came out above the lawns."
TENNYSON'S Princess.

THERE are few rock substances on the surface of the globe which have received more discussion and been more investigated than myself. I am somewhat proud of the attention I have received in this respect, for most of the leading geologists of every country, during the last century, have devoted themselves to the task of seeking out my antecedents. I am acquainted with a whole library of books, all most learnedly written, and various of them proving the reverse of the other, which have been penned on this inexhaustible subject. Even yet the question can hardly be regarded as finally settled. Every now and then some moot point or another crops up to engage the attention of philosophers, but, thanks to the progress of other sciences, the investigation of these is no longer confined to verbal expressions. It is not a little amusing to remember the hot discussions which were held over me at the beginning of the present century. Philosophers though they professed to be, the disputants resembled political squabblers more than anything else. One set declared I was born amid fire; the other that I was of purely watery origin. Each party believed in their own *ipse dixit*, and, as nothing could be absolutely proved, backed their own opinions by personalities. Somehow or other the former sect, who were called Plutonists, got the better of the latter, who were termed Neptunists. (The origin of these phrases my listeners will not find it difficult to understand.) But my Plutonic commentators carried their victory too far. Not content with proving that I was not a mere aqueous rock, they proceeded to declare I was nothing more nor less than one which had cooled down from a fused condition, something like iron slag; nay, it was even urged that I was older than any other rock, and the theorists mapped out an idea—which existed for many years afterwards, chiefly owing to its remarkable novelty—showing how the whole universe was formerly one great cosmical fog; that this diffused matter was condensed into suns, planets, and satellites, each of which existed for ages in a molten condition, owing to the heat evolved during the process of condensation; that the exterior of each planet cooled during the time which followed, and that granite formed part, or whole of this cooled envelope! Such in brief was the orthodox notion of my birth, little more than a quarter of a century ago.

Shall I enlighten my readers a little as to the nature of my mineralogical composition? I feel sure that most of them are acquainted with it already, but, if only for form's sake, I must go through with it again. My name is of Latin derivation, and was given me on account of the granular character presented by my different minerals. Generally speaking, these are four in number—Quartz, Felspar, Mica, and Hornblende. Very frequently there are also traces of other minerals; but these are the commonest, and in fact those which make up my bulk. The *Quartz* portion you may tell by its glassy appearance, and usually milk-white colour; whilst another good test is its superior hardness. This mineral is almost pure silica, and is one of the most refrangible of known substances. It can with difficulty be slightly dissolved in hot water, under great pressure; whilst it requires a great deal of heat to melt it, and, generally speaking, some sort of *flux* to set it a-going. The next most abundant mineral in the constitution of myself and relatives (for our name is Legion) is that called *Felspar*. Your eye may detect it in my mass, by its pink or flesh-colour, whilst it is so soft that you may scratch it with your finger nail. It is owing to the unusual abundance of this mineral that I am sometimes so friable or "rotten," as the felspar decomposes and then causes the other minerals to fall asunder, just as the bricks of a wall would if all the cementing mortar were to decompose away. In many districts, as in Cornwall, where granite comes to the surface and has been subjected to atmospherical wear-and-tear for ages, it is not uncommon to find the fine felspar wasted into a newer deposit. Such is the well-known "kaolin," or China clay of commerce. The chemical composition of felspar is more complex than that of quartz. For instance, although its commonest elements are silica and alumina,—the former the base of common sand, and the latter of clay,—there are also contained in it more or less of soda and potash, lime, magnesia, and iron. *Mica*, the next commonest mineral I possess, is so well known as hardly to need description. All my listeners are surely familiar with the small, thin, silvery-looking scales contained in almost every piece of granite. Its ingredients are much like those of felspar, only differently mixed. Frequently *Hornblende* is a mineral entering into my composition, when you will readily recognise it from its black or dark olive-green colour. When it is very abundant, it produces a rock varying from dark grey to black. A great number of what may be termed varieties of hornblende are known to mineralogists. Its chemical composition, generally speaking, is about one-half silica, more than a quarter magnesia, and little more than half a quarter lime: besides these, there are usually traces of iron, alumina, and fluoric acid.

I mentioned above that I had many relatives, who were more or less nearly connected (I cannot say by blood, but by mineralogical similarity of composition). These take various names, on account of their leading peculiarities. Among them the commonest is *Porphyry*, which takes its name from the purple variety used by the ancients in making vases, &c. This you may know from the large and distinct crystals, usually of felspar or quartz, which are imbedded in the granular matrix. Through porphyry granite passes into all sorts of allied igneous rocks, such as *Claystone-Porphyry*, *Clinkstone-Porphyry*, *Felspar-Porphyry*, and so on. When hornblende takes the place of mica in the composition of granite, the latter goes by the name of Syenite; when *talc* supplants mica, the result is called *Protogine*. A fine-grained compound of felspar and granite, with equally minute scales of mica, gives to you the varietal name of *Pegmatite*. According to the number of minerals entering into our composition, I and my relatives are roughly classed as *Binary*, *Ternary*, and *Quaternary* granites. All this detail of structure may sound very dry and tedious; but it is absolutely necessary to go through with it, if my listeners wish to be more intimate with me.

Although I have not a distinct recollection of my birth (as indeed, who has?), yet I have more than a suspicion that such elements as soda, potash, lime, &c., greatly assisted as fluxes in bringing me into my original molten condition. I have mentioned the great number of relatives who claim near or distant kinship with me, and I have now only to remark that their affinity to myself has been determined solely by the different circumstances attending their origin. I distinctly and utterly refute the idea that the first-formed crust of the globe was a granitic one! I am fully persuaded it could not possibly have been granite, and I will give you my reasons by-and-by for this seemingly bold assertion. What that cooled crust was, I doubt if science will ever be able to discover. But the fact that it was not granite does not in the least invalidate the theory that every sun, planet, and satellite was condensed from nebulous matter. This theory must rest on other grounds, and, singularly enough, additional facts are coming to its support every day. Men have not the slightest idea of what the *primitive* rock or crust of the globe was. The antiquated notion that it must have been granitic arose out of mistaken associations. It was found that, however old might be a stratified rock, whether containing fossils or not, some variety or another of granite was older still. Hence followed the hasty deduction, that originally one granitic crust encircled the fluid matter of the interior of the earth. It was thought that subsequent rocks were themselves formed out of the wear-and-tear of this granite, that the latter was in many places covered up by its own débris, and that the so-called metamorphic rocks were those first formed as stratified deposits, but altered to their present appearance through the intense heat of the newly-created seas, along whose bottoms they had been elaborated!

All this is wrong, and it behoves me now to descend from the region of pure hypothesis to that of fact. It is just possible, speaking generally of all the varieties of my family, that *Protogine* may be oldest. This, however, has never been thoroughly determined. One of my reasons for believing I could not have required any very great heat to reduce me to the molten condition, and that in this process the agency of water, as well as of heat, was necessary, is as follows:—Many of the larger quartz crystals entering into my composition are hollow. Frequently these hollows are more or less filled with water. Now it is a known fact that molten matter at a white heat requires its temperature to be considerably lowered before it can even evaporate the water mechanically mixed with it. It has been recently shown that crystallized matter which has undergone pure igneous fusion, has usually cavities in its crystals, not containing water, but either stony matter or a kind of glass, and, in many cases, even a perfect vacuum. Hence the conclusion is arrived at that in the case of coarse-grained granite, containing much quartz, there is actually more proof of the action of water than of dry, igneous fusion. It is more than probable, therefore, that pressure, heat, and water combined, in the deeply-seated parts of the earth's crust, would cause the rocks to be reduced to a kind of paste, and that

Fig. 1.

Microscopic Section of Pitchstone, showing dendritic crystals.

this paste, cooled under such circumstances, would be some variety of granite. I can hardly enter into the abstruse details of the deductions which have been made from the chemical and microscopical examinations of myself and relatives. Suffice it to say they result in proving that *pressure*, and this, generally speaking, of overlying rocks stratified or otherwise, is a preliminary and indispensable necessity to the formation of granite; that, if pressure be absent or less than that required, notwithstanding all the other requirements may be present—such as heat, similarity of mineral ingredients, &c.

Fig. 2.

Microscopic Section of artificial Porphyrine, showing ditto.

—such a resulting igneous rock would *not* be granite! It might be a variety of porphyry, or basalt, or greenstone, or, if all pressure were removed, and the molten matter allowed to cool in the open air, simply ordinary *Lava!* From a microscopical examination of various granites, it has been shown that those of the Highlands of Scotland indicate their having been formed under no less a pressure than twenty-six thousand feet of overlying rocks more than were the granites of Cornwall. There is good reason for believing the latter to have required at least forty thousand feet of rock-pressure; so, in that case, the granites of the Highlands must have been formed when sixty-six thousand feet of overlying rocks were piled above them!

One is naturally astounded by the magnitude of these operations, but I assure you there is little doubt as to the general correctness of the deductions. In this way the mineralogical construction of myself and others supplements the teaching of organic remains, as to the immense antiquity of the globe! Nothing short of an eternity of time would have sufficed for all the changes which have been rung upon it. There is reason to believe that many of my granitic relations are nothing more or less than *re-melted* stratified rocks, with their enclosed fossils! As these rocks have been slowly depressed or submerged, so as to bring the lowest-seated portions within the influence of the earth's internal heat, they have been first metamorphosed into a similar condition to gneiss and mica-schist, and, if the sinking went on, have passed through this stage into that pasty condition which deprived them of all stratified structure, and converted them into what I am myself! Then succeeded a reversal of the movement; so that this granite would be thrust slowly upwards with all the overlying strata piled above it. The movement went on until these were tilted into a continuous mountain-chain, or high and extensive table-lands. Meantime the granite nucleus would form the heart of such mountains, the strata dipping away on each side, as in the Himalayas.

I fancy I hear some of my listeners remarking— "But if granite can only be formed under such immense pressure, how is it we find such large areas of country where nothing else is to be seen?" In the answer to this we have the gist of the argument, and I would respectfully ask the special attention of my audience to it. Let them ask themselves where the materials came from to form the Laurentian, Cambrian, Silurian, Devonian, and, in short, all the other subsequent formations? They could only have been formed out of the waste of still *older* and already solidified rocks. Each formation, therefore, represents the amount of wear-and-tear which went on during the period when it was deposited. If there had been no compensation against this levelling process, all the highest grounds would soon have been worn down to a common level, and the elaboration of more recent deposits been self-checked. But each succeeding formation shows that this was not the case, and indicates that the physical arrangements of our planet have been much the same through all time to what they are at present; that atmospherical and marine wear-and-tear were counterbalanced by upheaval from beneath; that the external force emanating from the sun and resulting in all these atmospherical effects, was exactly adjusted by the native force of the earth, exerted from the interior outwards. These two have exactly checked each other from the beginning, otherwise the great life-scheme of our globe would never have had time for its development!

I hope I have been successful in explaining a great

Fig. 3.

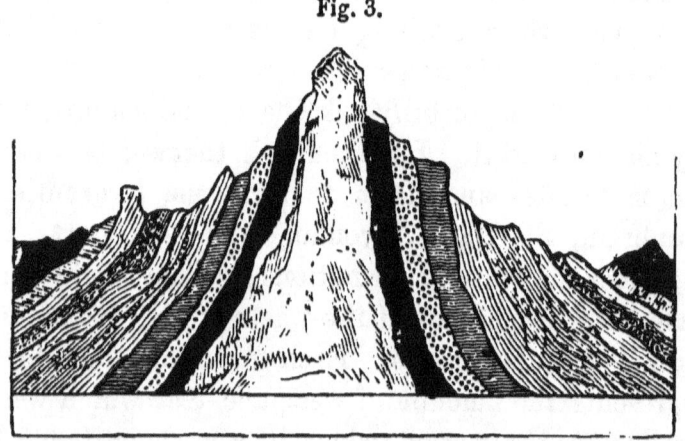

Section showing Granite nucleus, with strata lying on its flanks, the overlapping and continuous portions of which have been denuded off.

difficulty, and that my listeners now see the reason why I and my relatives come to the surface. It is *because the rocks which overlay me at my birth have since been stripped off, and slowly removed by atmospheric and other agencies.* All the formations which were then piled above me, are to be found in stratified rocks of later date; therefore, the period of my birth is not limited to any particular geological epoch. I am found at the surface, surrounded by rocks of every age, even including those of the Tertiary. Wherever the pent-up force of the earth's interior has thrust us up, there have we slowly elevated the rocks lying upon us. In many cases this elevation has been so slow that it has hardly exceeded the rapidity with which these overlying rocks have been gradually worn away! Think of the vast antiquity of the earth's crust, as indicated by these facts alone! Since the granites of the Highlands of Scotland were formed, twelve miles of overlying material must have been removed! Where has it all gone to? Ask the nineteen miles in thickness of the known stratified rocks, all of which have probably been formed since that granite itself. You scarcely need be afraid of Time, when you have Eternity to draw upon!

CHAPTER II.

THE STORY OF A PIECE OF QUARTZ.

"God worketh slowly; and a thousand years
He takes to lift His hand off. Layer on layer
He made earth, fashioned it and hardened it
Into the great, bright, useful thing it is;
Its seas, life-crowded, and soul-hallowed lands
He girded with the girdle of the sun,
That set its bosom glowing like love's own
Breathless embrace, close-clinging as for life;—
Veined it with gold, and dusted it with gems,
Lined it with fire, and round its heart-fire bowed
Rock-ribs unbreakable; until at last
Earth took her shining station as a star,
In heaven's dark hall, high up the crowd of worlds."
BAILEY's Festus.

"FACT," they say, "is often stranger than fiction." I do not think you will find this old saw better illustrated in the whole series of geological teachings than in my own history. That history is connected with one of the grandest discoveries of late years, inasmuch as it carries back the antiquity of the globe even beyond the mighty ages which had already been claimed for it. Indeed, the practical effect of this is to show the geologist that *time*, as a factor, has nothing to do with his investigations. That simple *relation* in the succession of events is all he can safely arrive at; and that his finite mind can no more conceive of the myriads of years which are included in the world's biography, than it can sum up in human arithmetic the stars and systems which crowd the illimitable realms of space! Within the last ten years a clearer geological knowledge of my origin has caused geologists almost to double the already known antiquity of the earth. At the time I mention, or thereabout, it was usually understood that the *Cambrian* period was the oldest and most primeval. The human mind is essentially conservative, and although geologists reasonably claim to be more catholic than most men, they are under the same influences. This is indicated by their unwillingness to make the world appear *older* than they possibly could help. Hence such terms as "Primary," "Primordial," &c. applied to the ancient strata—which nevertheless are all much younger than myself—are so many landmarks which have shown this tendency in the human mind. It may be, that although the geological formation to which I belong is undoubtedly the oldest known at present, in any country, subsequent research may eventually make known an older period still. The difficulty in doing so, however, will be considerably heightened by the fact of all these oldest rocks having passed through many changes, by heat and chemical action, so that nearly all traces of their former fossils are effaced, and thus they are reduced to a similarity of mineral condition all the world over.

There are few of my readers who are not acquainted with my general appearance. They have gathered me as a milk-white pebble by the sea-beach, or have admired me as they climbed the Scotch mountains and saw me sticking out of the contorted rocks like a huge white rib. Or, they may have been more pleased still with the geometrical shapes which my substance is capable of assuming as a six-sided, pointed crystal. It is of my former state, rather than of my latter, that I intend now more particularly to speak. And yet it is necessary for me to say that there are two common conditions in which I am usually to be found. One is as *Quartz*, the other as *Quartzite*. These terms are merely significant of appearance, and include little or nothing of chemical difference. *Quartz* proper is usually found in veins, having been forced into fissures when it was in a soft, heated condition. *Quartzite* has not so completely lost all its original structure, and its particles or grains may often be seen retaining their original water-worn form. Again, *Quartzite* does not occur as an intrusive rock, but in huge stratified masses, hundreds of feet in thickness. And yet you may find transitions in these two extreme states of my family—even from the transparent crystal condition of the " Brazilian pebbles " to the coarse-grained and resinous appearance of quartzite.

Let me be thoroughly understood. Although I am representing that great, and at present *oldest* epoch in our planet's history—the *Laurentian*—I should not like you to fall into the mistake of supposing that I am limited to it alone. On the contrary, formations of much more modern date than that to which I belong are rich in quartz veins and even beds. In short, any rock that has been exposed to the same influences as myself, if it contained the same chemical substances, would also become quartz as the result. They tell me that I am chemically composed of only one substance—*Silica*. My normal condition is transparent and colourless, although I am rarely found like this except when in geometrically-shaped crystals. A milk-white colour is that which I commonly affect, and this is due solely to the rate at which my parent mass cooled down. Hence it is that geologists can more or less tell from my appearance the circumstances which attended my birth. From the pure, transparent condition I mentioned above, I pass through a great many modifications, and in each stage of these I am known by different names. But with the exception of very slight mixtures of other ingredients than this silica, I continue the same throughout; thus, when I am of a violet tint I am called *Amethyst;* when of the colour of sherry, *Topaz;* when of a smoke-brown hue, *Cairngorm*, &c. Mixed with other chemical substances I pass into *jasper, flint, chalcedony, agates*, &c., in all of which you will find that the largest portion of their whole bulk is silica.

C

Up to the time when the geological formation to which I belong had been discovered, as I before remarked, the *Cambrian* was looked upon as the oldest. But there were a series of schists, quartzose rocks, &c., which were still older than these, and which usually went by the name of *Metamorphic*, or " altered " rocks; thus committing them to no particular geological age. By many these rocks were regarded as *transitional*,—that is, as passing from an igneous to a stratified condition. When it was imagined that all the granite rocks were formed as the outer crust of a once molten globe, then, it was also thought that the rocks which were formed along the bottoms of the hot seas must be of a very peculiar character. In short, these mica-schist, quartz, and gneissose strata were regarded as having been deposited and solidified under such circumstances. Their absence of fossils, and proofs of having experienced great heat, were thought to bear out this view. I hardly need tell you how erroneous it was. The Cambrian period was believed to be that when Life first appeared on the Globe. Now this supposition is known to be as wrong as that which accounted for the mineralogical appearances of the metamorphic rocks.

Although I am speaking only as a humble piece of quartz. you must remember that, when I am narrating the circumstances of my life, I am at the same time giving those of the mica-schist, gneiss, and altered limestones, which, equally with myself, belong to the Laurentian epoch. Indeed, the last-named rock, greatly altered though it is in appearance, so as to resemble loaf-sugar, could, perhaps, tell you more of the vital conditions of the ancient Laurentian seas than I can. First, let me impress you with the fact that when we were formed, collectively, we did not differ in appearance from the sandstones, clays, and limestones of either the present or of any bygone geological era. All this wonderful alteration in our appearance and structure is due solely to the subsequent changes we underwent. Of these I shall speak presently. If you know anything of the great deductions of geology, you will be aware that the farther you go back in time, the fewer and simpler are the forms of life which inhabited the earth. It was the general poverty of species, accompanied by their lowly organization, which caused the Cambrian epoch to be regarded as the first platform of Life. Now when you go farther back in time, to my own age you will find that the organisms are still lowlier. Indeed, of the objects that lived in the seas where I was originally deposited as a thick sheet of ordinary sand, all that I can remember is one abundant organism now known as *Eozoon*, or the " dawn-animalcule," in allusion to its primeval antiquity, It was lowly enough organized, being little above the natural history rank of the common sponge. This marine creature lived on the sea-bottom in vast quantities, and there grew by the addition of layer on layer of younger forms, just, as I am told, is the way in which coral reefs grow in modern seas! Like the latter, it absorbed its carbonate of lime from the sea-water, and thus caused great masses slowly to accumulate. This was in the deeper parts of the sea, where the water was clear, and free from muddy sediment. But my recollection goes no farther to any animal type. No fishes swam in the blue water; no crustacean crawled over where I lay! Occasionally the rivers brought some lowly-organized vegetables in entangled masses, or sea-weeds drifted into my neighbourhood, and eventually became entombed in the sandy mud—my then condition. An impure coal was thus formed, and when the rocks underwent their great transformation by the agency of heat, this vegetation somehow or another passed into *Plumbago*, or "black-lead," as it is commonly and erroneously called. The great amount of carbon —more than there is in many kinds of actual coal— which makes up the composition of plumbago, had long indicated its vegetable origin. How lowly organized were the land plants of the Laurentian period you may guess at from the fact that many ages afterwards, during the Carboniferous epoch, they existed only as gigantic club-mosses! What I have said about the vegetable origin of "black-lead" applies as logically to the origin of the Laurentian limestones. Some of the beds are as much as fifteen hundred feet in thickness, but altered throughout. As geologists are now aware, the limestones in every other formation are always of *vital* origin—that is, they have been formed by the accumulation of coral reefs, shells, &c., cemented, perhaps, by a still greater bulk of microscopic organisms. The white chalk of Norfolk is nearly as thick as one of these beds of Laurentian limestones, and yet, to the naked eye, it offers no explanation of its origin. It is not until you have applied the microscope that you perceive it to be almost entirely built up of the shields of animalcula, some of them of the same species as are still living in the Atlantic! If, therefore, the limestones of every known geological period have been formed by *vital* agency, one would imagine that those limestones, whose organic remains had been obliterated by the great heat to which they have been subjected, might be reasonably put down to the same origin. Again, the various *phosphates*, &c., found in these altered limestones, plainly tell of animal life having been employed in elaborating them. But, mighty though the transitions have been through which the whole of the Laurentian rocks have passed, *all* traces of fossils have not been lost. The limestones yet contain myriads of *Eozoa*, as plainly showing they were formed by its agency, as a coral reef tells you how its bulk grew to its present size.

Twenty thousand feet of material had been strewn along the bottoms of the Laurentian seas in various places, the material varying according to its neighbourhood to the mouths of rivers, &c., whence it was brought. The solidification of this mass took place at the same time as its deposition. A great plutonic change then occurred, and what had been sea-bottom for ages, eventually became dry land. Then followed a period of submergence, when it was once more sea-bottom, and had piled over it ten thousand feet of extra material! You ask how I know all this, and I reply by pointing to you how the upper ten thousand feet of rock lie *unconformably* on the lower masses. By "unconformability" I mean that the dip of their beds is not the same, the lower being different from the upper. This plainly shows that the lower beds were uptilted before the upper were formed, and that both series partook of the movement which finally elevated the upper Laurentian beds into dry land, in which state they remained during the subsequent Cambrian epoch.

You can readily understand how the Laurentian rocks, being the first formed, must have undergone more changes than any other, inasmuch as they have had to partake of all that has gone on since they originated. It is a wonder that we now find any of them uncovered by rocks of subsequent date; nor should we, had it not been for those great atmospherical denudations which have stripped off miles in thickness of overlying rocks, so as to expose those of an older date. The Laurentian strata have had, perhaps, miles in thickness of the rocks of other formations piled above them. They have had to undergo those great depressions which eventually brought them so much under the influence of the earth's internal heat. Masses of granite, trap, porphyry, &c., have been intruded through them, and thus they have been squeezed and contorted in the most fantastic manner. The sandstones, some of them five hundred feet in thickness, have been so affected by heat as to become *quartz*, or quartzite. Here, then, you have the secret of my origin—the whole history of the changes which brought about my present appearance! The limestones that were contemporaneous with myself were altered so as to resemble loaf-sugar, and had all, or nearly all, their organic remains obliterated. The shales and slates became transformed by heat, chemical change, and pressure, into *mica-schists, gneiss, felstones,* &c. So that the very peculiarity in dip, contortion, absence of fossils, and mineralogical changes, which mark all the rocks of the Laurentian age, tell of their vast antiquity; whilst the similarity in composition of these rocks in all parts of the world,—in Ireland, Scotland, and North America, as well as the prevalence of similar lowly-organized fossils in their limestones, indicate they have passed through the same transformations since they were contemporaneously deposited as limy muds, sands, and clays along the floors of the primeval seas!

CHAPTER III.

THE STORY OF A PIECE OF SLATE.

"It is a lonely place, and at the side
  Rises a mountain rock in rugged pride;
  And in that rock are shapes of shells, and forms
  Of creatures in old worlds, and nameless worms—
  Whole generations lived and died, ere man
  A worm of other class, to crawl began."

CRABBE.

WAS not always what you now see me. Far, far back in that almost infinite past, which geology claims before it can explain its phenomena, I was lying along the bottom of a tolerably shallow sea, as part of an extended sheet of fine mud. My birthplace is registered in the heart of the North Welsh mountains, and the formation to which I belong goes by the name of the Cambrian.

Its rocks form some of the grandest scenery in the world. Steep precipices, on which grow rare ferns and wild plants, frequently too tempting to the botanical student, are the result of succeeding dislocations, jointings, and bedding. Mountain streams brawl over them; and waterfalls, whose substance is evaporated into prismatic mists, pitch from the precipices of these Cambrian hills. Frequently the rocks are so hard and bare, that even the lichen and moss fail to obtain foothold, and so the naked slate shines in the varying sunlight in coloured shades from pink to deep blue. Here, with the gathering cumuli, ring-like crowning their peaks, the Welsh hills stand forth in all their characteristic grandeur. No wonder that crowds of tourists should strive to forget the cares of business, and endeavour to get a mouthful of purer air, whilst climbing their steep sides!

It requires some faith in geology to carry the mind definitely backwards to the time when these rugged hills were extended sheets of marine mud! But no mathematical deduction is more certain. You never find clay or sandstone rocks so full of fossils as limestones, for the simple reason that the former are of *mechanical* origin, and the occurrence of organic remains is therefore accidental. Whereas limestones are of *vital* origin, resulting from organic agencies almost entirely.

You examine the slate rocks of which I am a humble representative. Their colour and general texture you easily recognize from the too familiar appearance of the London housetops. But, when in position, you are scarcely prepared to find that what you had imagined to be the result of bedding or lamination in the slates is actually due to what is termed *cleavage*. This is a peculiar feature about thin-bedded, argillaceous or clayey rocks, that they undergo, when subjected to pressure, and perhaps heat as well, a certain change, which is in reality a sort of rude, massive crystallization. By virtue of this process, the rock splits not so readily along the lines of stratification or bedding as along that of the cleavage, or planes of sub-crystallization.

In addition to this structure, which is frequently diagonally across the line of stratification, these slate rocks are broken up into large cubic masses, caused by great joints traversing the rocks, irrespective of any previous alterations.

Fig. 4.

Showing foldings in strata.

The stratification itself is not horizontal, but frequently pitched up at a very steep angle, and commonly the rocks are contorted into a series of ribbon-like foldings. After all this cleavage, jointing, dislocation, and faulting, the solid rocks have been subjected to thousands of centuries of atmospheric and marine wear-and-tear! Can it be wondered at, therefore, that there should result from all these combined agencies, continued through untold millenniums, all that wildness and grandeur of physical scenery which distinguish these old Cambrian rocks wherever they are met with?

THE STORY OF A PIECE OF SLATE. 27

The old rocks, especially those of an argillaceous character, are nearly always marked by contortions, to which those of a later date are strangers. It is from amidst them also that we have great bosses of granite coming to the surface, the contorted slate rocks surrounding them on every side. How is this? I will endeavour to explain.

My hot-tempered friend, the piece of granite, told you how it was absolutely necessary to his origin that the molten rock of which he was portion should be overtopped by a tremendous thickness of material when it was cooling. This my own experience will bear out. The contortions which characterize my family equally required an amount of overlying material to be piled upon them, or they could not have arrived at such singular appearances.

A mass of half-hardened rock, if displaced by a foreign body, such as a boss of granite being thrust up, would rise up as one great hill or mountain. But if there was sufficient pressure overlying the formation thus disturbed, then it would be thrown into a series of foldings, in order to make place for the laterally-intruded material. Of course the whole exterior surface would then be elevated; but this elevation would not be in a conical form, but along a large tract of country.

In geological books you will find how, on a small scale, this experiment has been conducted. A series of layers of cloth has been formed; pressure was applied to the sides, when the surface naturally rose into a sort of mound; but the moment a heavy weight was laid on the top cloth (thus representing the overlying material of which I spoke), then the layers of cloth, when pressed at the sides, became folded up into a series of contortions. My hearers will now see why granite outcrops should frequently be the companions of slaty contortions; for the agency of overlying rock-masses, which originated the former, by their pressing weight caused the latter, when disturbed, to assume the wrinkled, fantastic shapes they now present!

It is not long since the Cambrian formation was deemed the oldest in the world; even its most learned and indefatigable observer called it the *Protozoic*, imagining its organic remains to be the " first life-forms." This provisional place of honour, however, has since been bestowed on a still older, and of course even a more contorted and metamorphosed class of rocks, termed *Laurentian*. Whether this in its turn will have to give place to one older still I cannot tell; but this I know, that the more you study the rocks and their contained fossils in the field, the more will you be convinced of the enormous antiquity of the earth, and of the incalculable period during which life has been divinely manifested upon it! Human arithmetic will never be able to compute my own age, and therefore the very attempt would be futile. Seeing that we slate rocks are, as far as England is concerned, the oldest known, who can wonder we should be found in such a dislocated and contorted condition? Have we not had to bear the heat and burden of the day? All the rocks of later date have been uplifted into dry land from the sea-bottoms on which they were formed; and seeing we were older, it was impossible to elevate them without also raising us at the same time; so that the alternate elevations and depressions to which we have been subjected are innumerable. Meantime the overlying formations have been slowly eaten away, attacked either by atmospherical forces or by marine denudation.

Far distant though the period of my birth may be, I have a lively recollection thereof. I am well provided with "hints to memory," in the shape of fossils impressed on, or included in, my parent bulk. I have only to turn to these, and immediately the old life-scene vividly recurs to me. What a strange time it was, and how different to anything I have since beheld! I can readily understand how the earlier geologists should reverently regard our fossils as the first created. In them Nature seems almost to have "tried her 'prentis han';" for these earlier organisms bear about them the impress of a lowlier fauna. Not that any are found which cannot be referred to existing natural-history orders, for Nature, like her Lord, knows "no variableness, or shadow of turning." Her plan has been to fill up the outline, and this has been slowly consummating during the unknown ages which have elapsed since the Cambrian period. Hence it is that the further you go back in time, the more simple is the *facies*, or general appearance, both of animals and plants. It is possible that, at the time I was born, the dry land was sparsely covered with a humble flora; but it will be evident that as I am of purely marine origin, I cannot speak with certainty of what took place elsewhere. I have a dim recollection, however, of certain obscure mosses, lichens, and perhaps reeds, but nothing more certain. That there was dry land, and that this dry land was watered by extensive rivers, I have not the slightest doubt. Otherwise, where would the materials have been derived which make up the bulk of my parent formation? And, that this material was *slowly*, and not rapidly obtained, you yourselves may easily see from the fineness of the particles which enter into my composition. For the Cambrian formation is no less than eighteen thousand feet in thickness; and, with the exception of certain beds in the middle of this immense bulk (called by geologists respectively Harlech grits and Lingula flags) the rocks of this period are principally fine-grained slates. Even the grit-stones and flag-stones afore-

then, that the strata of a formation nearly three and a half miles in thickness could be deposited in only a tolerably shallow sea? The question is natural enough, and I reply by stating that whilst these strata were slowly forming, the sea-bottom was as slowly subsiding. Hence it remained at almost the same depth during the long period when these fine muds were thrown down. You will find my statement verified by the fact that in the Lower Cambrian (in a group called the Longmynds) the tracks, holes, &c., of marine worms (termed *Arencolites*) are found distributed through a vertical thickness of over a mile of rock. Nor are these humble organic remains scarce; they occur in countless myriads. After the deposition of the Lower Cambrian rocks, as far as I can recollect, the sea began to get deeper; the deposits formed along its bottom did not quite equal the rate of depression, and so the depth of water increased; but before then I well remember how comparatively shallow the sea was. This is attested not only by the countless fossil worms which have won a geological immortality from the trails they left on these early sea-bottoms; but also from the ripple-marks which equally characterize the same set of strata. Nay, we have even evidence of extensive mud-flats, for many of the beds are pitted with rain-drops, and marked with sun-cracks. Thus, far back as English geology can take you, you have evidence of exactly the same kind of meteorological agencies as those which now regulate the physical well-being of the external globe. Cloud and sunshine are testified to by these sun-cracks and ripple-marks. Vapours were raised by solar heat then as now, and the "bow was set in the cloud," although not as yet selected as a covenant to man!

In the same beds as these ripple-marks, sun-cracks, rain-pittings, and worm-tracks, we have innumerable remains of a small crustacean (*Palæopyge*), which used to flit through the shallow water in dense shoals. A pretty little zoophyte (*Oldhamia*) lived in quiet, sheltered spots, where it luxuriated abundantly, its little branched stems forming miniature forests along the old sea-bottom. These lowly creatures are almost all I remember of what is called the Lower Cambrian formation. The upper portion, however, is much richer in fossils; and well do I remember when these now petrified organisms enjoyed the pleasures of animal life. Between the deposition of the strata of these upper and lower formations there was a break in the locality where I was born. Probably somewhere else in the globe there will be found a formation (possibly limestone) which was elaborated during this provisional rest. Of that, however, I can only conjecture. Concerning the animals which lived in the Upper Cambrian seas, I can speak more positively. They were, first of all, far more abundant, both in species and individuals. Thus the basement rocks of this subdivision go by the name of *Lingula* flags, from the vast quantities of the fossil of that name occurring in them. The *Lingula* was a mollusk occupying the lowest class among shell-fish, that termed *Brachiopodous*, or "arm-footed," from the peculiar arrangement of the breathing and locomotive organs. Strange enough, this genus is still in existence, and you can hardly tell the difference between the horny shells of the living species and those which lived at this early epoch. Talk about genealogy; no other family, except that of the marine worms, can claim an antiquity so vast. Notwithstanding all the mutations through which the surface of our old world has passed—the upheaval of sea-bottoms into mountain-heights, the depression of mountains into sea-bottoms—this one genus of shell-fish has triumphantly survived them all! It is now, I am told, fast passing into extinction, the final lot to which so many genera of subsequent date have succumbed. Among other animals which lived at the time was a species of shrimp (*Hymenocaris*), whose remains may be met with in the same rocks. Along this sea-bottom, in various places, lived colonies of a kind of sea-lily, or rather, of an animal halfway between them and the more recent sea-urchins: these now go by the name of *Cystideans*. Furnished with a short footstalk, which served to anchor them to their selected habitats, they flourished on the lower forms of life which swarmed in the waters of these primeval seas.

Later on was introduced a crustacean afterwards

to become famous, both for its abundance and the number of generic and specific forms it assumed. This was the well-known *Trilobite*. Several genera, and still more numerous species, were in existence, and so fast did the newly-introduced species breed, that they soon became the chief inhabitants of these early seas.

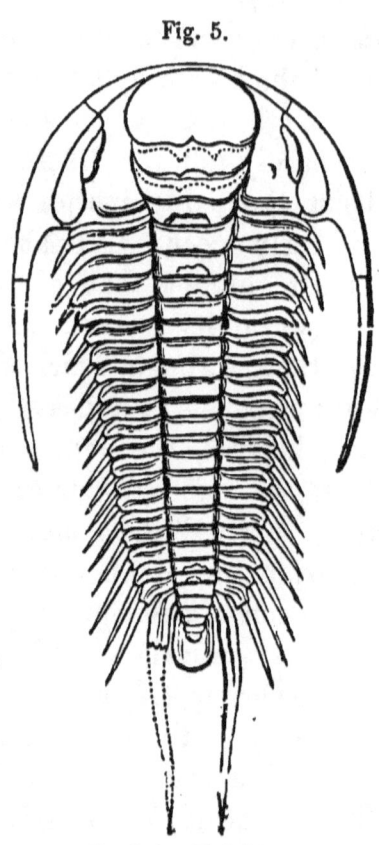

Fig. 5.

Cambrian Trilobite
(*Paradoxides Bohemicus*).

Most of my listeners are acquainted with their tri-lobed forms (whence their name), and have admired the jointed coat of mail which protected them, and, at the same time, gave them all the necessary flexibility for movement.

Out in the deeper water lived a peculiar kind of mollusk, whose type is still living. This is termed *Theca*, and its external protection consisted in a thin, almost glassy case; not so fragile, however, but that it has been carefully fossilized. But in gritty sandstones, or coarse slates, it is rare you will find any remains of the old calcareous shell of the various creatures I have named. Subsequent changes, most of all the percolation of draining water, have removed the limy material, so that the fossils found are principally as *casts*. Perhaps the lime thus removed has, in many cases, served as a natural cement to the sandy or clayey particles, so that much of the hardness which now characterizes these rocks may be originally due to the limy substance of the Cambrian inhabitants.

Towards the close of this remarkable period, other forms of life appeared, the total number of genera and species considerably increasing. Shell-fish of a higher grade were introduced, until the highest type—the *Cephalopoda*—was brought on the stage of existence in the shape of *Orthoceratites*. These were allied to the living nautilus, only they had straight chambered shells, instead of coiled ones. Their arms, something like those of a cuttle-fish, extended out of the last, or body-chamber; and on these, with their shell inverted like a spire, the creature would occasionally crawl over the muddy sea-bottom, where I was slowly forming. Belly-footed mollusca (*Gasteropoda*), in the shape of a genus which has been extinct since the time of the coal formation, crawled about, their gracefully coiled shells being as beautiful as any of their recent representatives. Thus did the Cambrian period come to a final close.

Of course my listeners cannot expect one poor

memory accurately to remember all the types of life then existing! Suffice it to say that, compared with those of subsequent periods, they were few and of a much lowlier kind: numerical abundance of individuals made up for poverty of genera and species. It was the dawn of life—when organisms were in the cradle. Betwixt this and chaos was a great gulf fixed. The first outlines of that grand scheme which should ultimately link inorganic matter with spirit were then rudely sketched. Time was ordained for the sole purpose of filling them up, and, when the object is completed, time shall be no longer! Even since this distant period, life has progressed until it has reached its physiological maximum in man. But in him, I am told, appear the germs of a new spiritual life, whose development shall extend into the future, just as organic life has been developed in the past! Such are a few of the reminiscences of a piece of slate! Of the agencies which uplifted me into a mountain-ridge, which consolidated the fine mud where I was born into hard slate, I cannot tell. These are all included in those chemical and geological changes which took place after my birth. But, whilst I have thus endeavoured to administer to the intellectual curiosity of man, I cannot forget that it is to these subsequent alterations that I am what I am, and that I now assist in roofing in and protecting the latest introduction of nature in the form of man!

CHAPTER IV.

THE STORY OF A PIECE OF LIMESTONE.

"Millions on millions thus, from age to age,
With simplest skill and toil unweariable,
No moment and no movement unimproved,
Laid line on line, on terrace terrace spread,
To swell the heightening, brightening, gradual mound,
By marvellous structure climbing toward the day.
Each wrought alone, yet altogether wrought
Unconscious, not unworthy instruments,
By which a hand invisible was rearing
A new creation in the secret deep.
Omnipotence wrought in them, with them, by them;
Hence what Omnipotence alone could do,
Worms did. * * * * *
Slime their material, but the slime was turned
To adamant by their petrific touch;
Frail were their frames, ephemeral their lives—
Their masonry imperishable. All
Life's needful functions, food, exertion, rest,
By nice economy of Providence
Were ever ruled to carry on the work
Which out of water brought forth solid rock."
MONTGOMERY's *Pelican Island*.

AM elected as spokesman for a common and well-known mineral, which is abundant in every geological formation. Our age, therefore, varies as greatly as it is possible for mundane time to allow. Chemically, our composition is always pretty much the same, being merely carbonate of lime.

In all the rock formations we are further distinguished from the sandstones, shales, and conglomerates, by our being almost wholly of *vital* origin, that is, formed through the agency of living beings; whereas the other rocks I have mentioned are the result of *mechanical* forces, wearing down and triturating pre-existing rocks, and then re-depositing the *débris* along old sea-bottoms. In consequence of this difference, the geologist finds in us by far the greater number of those organic remains, especially of marine animals, by whose aid he is enabled to sketch forth the development of the world's great life-plan.

As a rule, all limestones have been deposited, as fine calcareous ooze, away out in deeper water; consequently the circumstances have been doubly favourable for the preservation of any animals which might have died and become entombed in this limy mud.

The more boisterous conditions which prevailed in the shallower waters, where coarse sands and conglomerates were formed, prohibited such favourable preservation. At the same time, with the exception of what are known as freshwater limestones (which bear a very small per-centage to the other rocks of the earth's crust), I must acknowledge that the sandstones afford most valuable evidence of the *terrestrial* animals. This, as might be expected, is mainly owing to the fact that the latter were formed nearer to the shore, so that carcasses of land animals accidently drowned or carried into the sea by rivers watering large islands or continents where they lived, would sink to the bottom, and be buried up in coast deposits; whilst the sandstone and shale formations testify to the long-continued wear-and tear of the solid land by meteorological agencies. Therefore, the limestones bear out the idea of our planet's antiquity, by suggesting the immense lapse of time which must have occurred whilst simple and lowly animal functions were elaborating the greater proportion of all the limestone rocks.

But I intend to let each of these speak for itself. They are of age, ask them! Each contains its own suite of organic remains, the extinct creatures which lived and died whilst the limestone mass was slowly accumulating as calcareous ooze. They are tombs of the forgotten dead—stony scrolls, written within and without.

I myself belong to that most interesting geological formation known as the Silurian. Away in the heart of the "Black Country," where no less than thirty feet of solid coal abut against their flanks, you may see cropping up an irregular and continuous ridge of limestone hills. It is thence I am derived. You may gather some idea of the forces which slowly upheaved these strata by seeing the steep angle at which they lie: a little more and they would have been quite perpendicular. But this

40   THE STORY OF A PIECE OF LIMESTONE.

upheaval was not violent or sudden; on the contrary, I distinctly remember its operating through

Fig. 6.   Fish, Trilobites, Brachiopods, Corals, and Graptolites of the Palæozoic epoch.

long-continued ages subsequent to the Silurian period. The process was so slow as to be almost imperceptible, for Nature knows little or nothing of those violent cataclysms which have been so foolishly ascribed to her! Examine the steep flanks of the Wren's Nest, near Dudley. There is hardly a space of a pin's point which is not occupied by the remains of some creature in which the breath of life was enjoyed countless millions of years ago!

You strike the solid rock with your hammer, and immediately the percussion liberates a heavy sulphuretted odour, which tells of the old animal oils in which the limestone is steeped. The very hardness of these rocks is more or less indebted to the same organic cause. I am told that when sculptors, now-a-days, wish to harden their plaster-of-Paris casts, they do so by boiling them in oil. The principle is the same with most limestone rocks of every age. They are steeped, saturated in animal oils; nay, in many places across the Atlantic, where these old Silurian limestones and shales lie so deep down as to be within the action of the earth's internal heat, these oils have been distilled out of the rocks, and have followed the ordinary habits of fluids. It is by sinking through the overlying masses that these oil-springs are reached, and the valued liquor comes bubbling to the surface. Well does it deserve its common name of *Petroleum* —" rock-oil."

But few people imagine, when its brilliant light is illuminating their comfortable homes, that they are indebted to distilled *Trilobites* for the luxury! Here is another form of that grand law of correlation of physical force. The ancient Silurian sunlight furnished the means of vitality to the creatures which then enjoyed life. It was stored up in their tissues, and given forth in their buoyant gambols and locomotive powers. And when they died, what remained in their diminutive bodies decomposed, passed into other chemical forms, was preserved until our own day when men unlock this ancient sunlight from its oleaginous condition, and turn it to direct heating and lighting account! Fancy sunlight bottled up in the form of trilobites and mollusca! No wonder these should present such stony and petrified appearances, when all the animal oils have been so completely drained out of them.

How long these Wenlock limestones (for that is the name by which this section of the Silurian formation is known),—how long, I say, it is since these limestones were upheaved and exposed to the action of the weather, I cannot say. Their hardness, as I have already mentioned, is most intense; but the wear-and-tear of the atmosphere has been such as to cause the fossils to stand out in relief; and a strange sight, therefore, is the exposed surface of the limestone slabs. The eye is bewildered by the number and variety of organic remains, each standing forth from the fine limy mud in which it was originally enclosed. Little or no vegetation grows on this bare limestone surface; the latter is too impenetrable to yield a foothold; and so the geologist has it all to himself. Heads and tails of *Trilo-*

bites, so plentifully dispersed that they immediately stamp the Silurian age of the rock, lie commingled with brachiopodous shells, worm-tubes, sea-mats, chain-corals, and encrinite stems. You require no prompter to remind you of the exuberance of animal marine life in this distant epoch, and yet the Silurian period immediately succeeds the Cambrian, about which my distant relative, the Piece of Slate, gave you an account some time ago.

Whilst the limy mud—which subsequently became hardened into solid rock, and then upheaved into its present condition—was being slowly formed in deeper water, nearer to the shore there were deposits of a different nature going on: these consisted of muds poured into the sea by rivers, or wasted by tidal and current action from old coast-lines; gradually, therefore, the limy deposits passed into the muddy ones, so that the line of junction was almost imperceptible. Occasionally the fine mud was carried further seawards than usual, and then a thin layer of argillaceous matter was thrown down over the limy material. This accounts for the frequent alternations of limestone bands and argillaceous shales which you have doubtless seen in every section of Silurian strata.

At various epochs during the immensely long period which elapsed whilst these beds were forming, alterations of the sea-bottom took place; the area where limy deposits had been forming became shallow, so that clay or mud began to accumulate over the same spot; or, the sea-bottom became deeper, and, in that case, calcareous or limy material slowly formed where mud had previously been accumulating. Occasionally, perhaps, the sea became so shallow that shingle-beds were strewn over the area where both lime and mud had been collecting. My hearers can readily understand operations like these; they are still going on over various parts of the earth's surface; but the time of observation has not been extensive enough to see what they can effect. Only that simple element of time is required —and our planet is changed as by the will of some powerful magician! And, for my own part, I do not see why the timid, unconceding spirits of modern times should begrude *time* to the geologist, any more than they do *distance* to the astronomer!

The various strata which vertically succeed each other in the Silurian formation plainly indicate the geographical changes which affected these ancient seas; and, at the same time, imply the vast lapse of time during which they were brought about. Suffice it to say, this Silurian formation, with its enclosed strata, attains a total thickness of no less than twenty-six thousand feet!

Leaving my junior brethren to speak for themselves when their turn comes, let me try and remember some of the physical circumstances which marked the epoch of my own birth. First of all, what a different geography marked the surface of the globe then from what there is at present! I believe there was a much wider extension of sea than there is even now, when it extends over more than two-thirds of the earth's surface; and, owing to there having been fewer disturbances at that time, the sea was more equable in depth; whilst, at the same time, the dry land was less distinguished by mountain-chains. In consequence of the equable depth (or nearly so) of the sea, and of the similar climature which the entire surface of the world seemed to have enjoyed alike, there was less difference in the animals and plants of various geographical zones; but this principle was in existence, although nothing like so broadly developed as at present.

The Silurian limestones of America, Asia, and Europe differ very little in their general *facies* of organic remains. You have no difficulty in recognizing the old features which struck you when examining the Dudley strata; but when more minutely studied, the naturalist makes out certain "colonies," caused doubtless by difference of geographical circumstances. As the time passed away during which the great sequence of beds belonging to the Silurian formation were being elaborated, other changes took place in organic life. The most marked feature was that of a progression from lower to higher types. Species multiplied, and the general total of life-forms became more varied and less cosmopolitan.

Fig. 7.

Lingula Lewisii.

The lowest beds of my parent formation go by the name of Llandeilo Flags, so named from the locality in North Wales where the typical section may be studied. They are, as their name implies, strata of flaggy sandstone, much worked for commercial purposes. There is a considerable quantity of limy matter in their composition, and this gives them a peculiar indurability. Interstratified with the beds of this deposit are immense layers of ancient volcanic matter,—basalts or tuffs: these

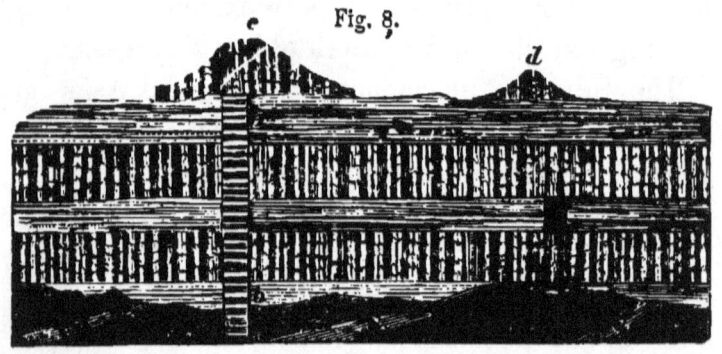

Basaltic rock interstratified with aqueous rocks; *b* Basaltic dyke.

flowed over the old sea-bottoms, when ejected from submarine volcanoes, or volcanoes situated near to the coast, as we find they usually are now-a-days. The ashes or tuffs were carried by the winds, and the ancient seas had their surfaces thickly strewn with cinders for hundreds of miles; these sank to the bottom, and alternated with the regular shore deposits. Succeeding the Llandeilo Flags, we have another division, known as the Bala Limestone, also named from a locality: it has interstratified with it beds of sandstone, slates, and volcanic rocks again, which were doubtless strewn over the old sea-bottoms just like those already mentioned. The Caradoc Sandstones, named from their locality in Shropshire, containing also shelly sandstones, with soft shales and conglomerates, lie above the Bala Limestone, and complete what geologists have termed the "Lower Silurian Rocks." They differ, as a whole, in Great Britain, from their comprising such a huge bulk of strata of igneous or volcanic origin. In some places these are actually thicker than the rocks of sedimentary origin. What a stormy, restless epoch was that! The old sea-bottom was subjected to shocks and volcanic overflow more intense than those in the neighbourhood of Iceland, where the Skaptar-jokul is quivering with suppressed rage and superfluous power! Then, again, these Lower Silurian rocks have neither so abundant, nor so highly organized a fauna as the rocks of later date.

Let me mention the next in order, before I give you my personal recollections of the extinct creatures you find imbedded in these rocks as fossils. The "Middle Silurian" strata commence with the Llandovery slates (another localism); after which you have the May Hill sandstones (about which not a few geologists quarrelled some years ago) and the Tarannon shales; altogether, this series is about two thousand feet in thickness, the Lower Silurian beds I have described being upwards of nineteen thousand feet thick. Next come the uppermost beds (to which I personally belong), known as the "Upper Silurians," and which attain a total vertical thickness of nearly five thousand feet. They include several deposits of minor importance; such as the Woolhope beds and the Wenlock limestones and shales, completing what is known as the "Wenlock Group." Then succeed the Ludlow beds, the Aymestry limestones, and the Downton sandstones, in the latter of which is found a bed composed of scarcely anything else than the bones, teeth, and scales of small fishes, belonging to the placoid and ganoid orders. It is in these soft shales you find the fossils so well preserved. The shells, although they have been extinct for unknown millions of years, still retain their beautiful iridescent nacre, which, however, soon decomposes by atmospherical influence.

Fig. 9.

Curved Orthoceras
(*Cyrtoceras Murchisoni*).

So much for the "stratigraphy" of this most interesting geological formation! At the forms of life which swarmed the seas of this distant epoch I cannot do more than merely glance. I have mentioned that, generally speaking, there was a progression. This is

THE STORY OF A PIECE OF LIMESTONE. 49

true only of the advance in the main, for, during the earlier portions of the Silurian period, huge *Orthoceratites* abounded, and these are among the highest classes of the mollusca. The muddy sea-bottoms swarmed with "sea-pens," now known as *Graptolites*, allied to the little Corallines so plentiful in modern seas. The chief difference between them being that the former were free and unattached, whereas the latter always adhere to some other body. But, of all forms of life, those of the Trilobite family were most abundant. Several hundred species are known to belong to the Silurian formation alone. They were crustaceans, allied to the King Crab* of the Moluccas, and at that time represented the lobsters and crabs of the present day. This is a group which has always been noted for its aberrant types. Like other crustacea, the Trilobites underwent metamorphoses or larval changes. So well do the old

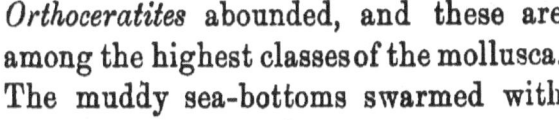

Fig. 10.

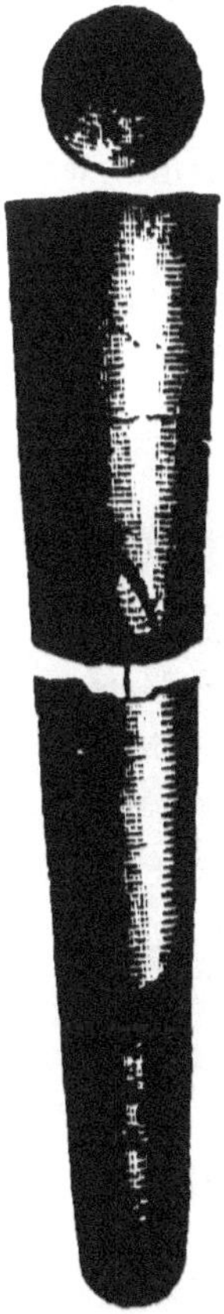

Orthoceras, upper part showing perforated chamber.

Fig. 11.

Sea-pen, or Graptolite.

* The larva or young of the King Crab very much resembles some of the ancient Trilobites.

E

rocks tell their story of ancient life, that the geologist has traced the metamorphoses of Trilobites through no less than twenty different stages, from the egg to the adult animal. In the last condition its body was enclosed in tri-lobed joints, which served as a defence, and at the same time were flexible enough to be adjusted to all the motions of their possessor. In fact, they served all the purposes

Fig. 12.

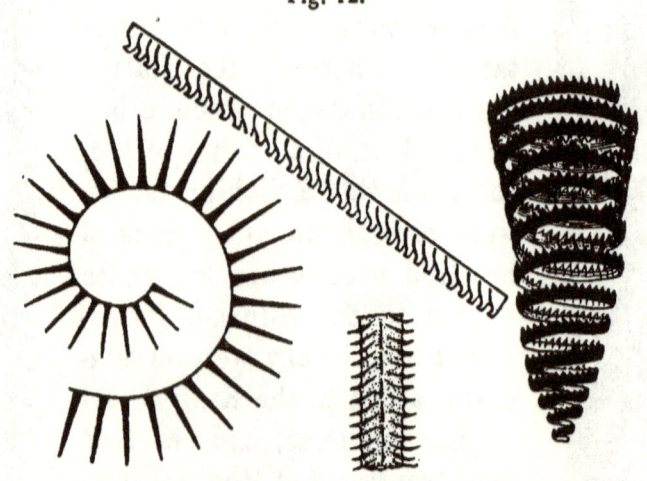

Various species of *Graptolites*, from the Silurian rocks.

of an ancient coat of mail. These various species of Trilobites literally swarmed in every sea of the Silurian period. There were species alike peculiar to deep water and to shallow, and the rocks formed under these different conditions (as I have above related) indicate which these species were. Well do I remember them crawling over the oozy sea-bottom, gorging the mud, as I am told earthworms now do, for the sake of the animalculous matter dispersed through it. Not long ago, some of these fossils were found which were supposed to have the legs attached to the under side. As a rule, however, the Trilobites are usually met with without these useful appendages, and no small discussion has arisen as to whether they had them or not—a discussion which is now set at rest. When any danger approached, they coiled themselves up like modern woodlice, and, in this state, you may not unfrequently find them fossilized. When the adult animal moulted, he did so at the junction of the head and carapace; and this accounts for the myriads of detached heads and tails found in every piece of Silurian limestone or shale. The Trilobite had compound eyes, arranged sessile, on half-round prominences, on which they were set like so many mounted jewels. Some species had not less than four hundred of these distinct eye-facets. Thus we find the structure of this little creature completely setting all those wild theories

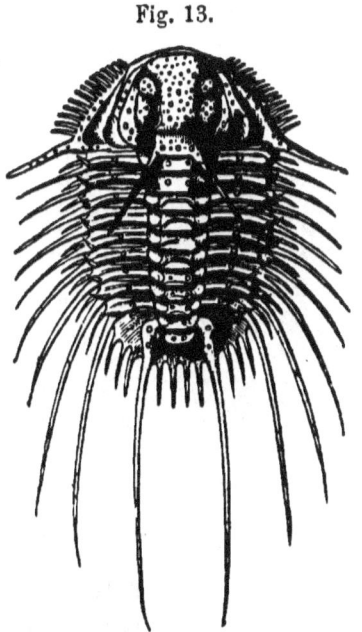

Fig. 13.

Spiny Trilobite (*Aciduspis Dufrenoyi*). Silurian rocks of Bohemia.

at defiance in which some people have indulged. Their eyes indicate a similar constitution of the atmosphere then to what it is now, for the passage and refraction of the rays of light. And this fact is supplemented by the sun-cracks, rain-drops, &c., which pit the sandstones, telling of meteorological action identical in its operation with the present. Indeed, all the facts go to prove that even at this distant epoch of the world's history, the light of the sun and the atmosphere of the earth were exactly like what they are at the present time.

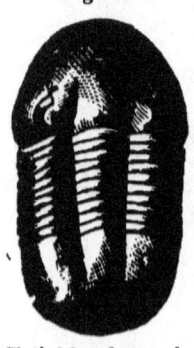

Fig. 14.

Trilobite from the Silurian rocks, Pultowa, Russia (*Illænus crassicauda*).

During the period of the "Middle Silurian" there was a great change in physical geography. How long a time had elapsed since the Lower Silurian strata had been formed, with their enclosed great sheets of volcanic lava and ash, may be guessed at from the fact that the May Hill conglomerates are composed of the waste fragments of the former; they had therefore been solidified into such rock as you now see them, and been uplifted from the sea-bottom into coast sections, and it was from their wear-and-tear, when in the latter condition, that the May Hill conglomerates were formed. Thus does the very structure of many of these deposits indicate the immense amount of time which elapsed during their elaboration. It was during the deposition of the

"Upper Silurian" beds, however, that life was most prolific—was most varied. The sea was aglow with huge coral reefs, around which swarmed sea-lilies, star-fish, mollusca of innumerable species, nautili, orthocerata (of whimsical and various shapes), and trilobites. The scene was most busy and most animated; the compound corals shone in various colours, and the adjacent sea-bottom was literally a submarine forest of crinoids, or sea-lilies. How abundant these lovely creatures were you may guess from the fact that you can scarcely pick up a fragment of Upper Silurian limestone without perceiving some of their detached ossicles, or jointed plates. In and out of these waving forests, with the arms of the animals representing branches, the innumerable species of trilobites swam, and crawled, and climbed. Every now and then some brightly-coloured *pecten* flitted past like a butterfly. Univalves (*Murchisonia* and *Euomphalus*) of delicate ornation and colour, slowly dragged their pretty shells about; the Cystideans, with their dwarfed stalks, but highly-ornamented and sculptured heads, dotted the sea-bottom. Over all, the occasional long arms of star-fish wound and unwound; delicately beautiful nautili, of various species,

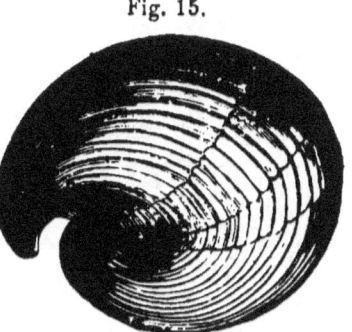

Fig. 15.

Pentamerus Knightii.

sometimes crawled, sometimes filled their air-tubes, and mounted to the surface of the water. The whole of Wenlock Edge, in Shropshire, is nothing less than an ancient Silurian coral reef, around which, millions of years ago, all the vital circumstances I have been attempting to describe took place! Of all these beautiful coral forms none were so lovely as the "Chain-coral" (*Halysites catenulatus*). Well does it deserve its name, for even now it appears like some watch-chain of exquisite workmanship interfolded in the solid rock! The largest of these corals was the *Favosites polymorpha*. Amidst all should not be forgotten the nests, groups, or even banks of *Terebratula, Atrypa, Rhynchonella, Spirifera, Producta, Strophomena*, and *Pentamerus*; all of them belonging to the lowest class of Mollusca, then in luxuriant abundance, but now waning into extinction. Towards the close of the Upper Silurian period, *Vertebrata*, in the form of *fishes*, made their appearance: at first they were few in number and small in size: but ere long they multiplied amazingly. They had their old feeding and breeding grounds, and along this part of the old sea-bottom their remains were of course most thickly accumulated. Such is the explanation of the

Fig 16.

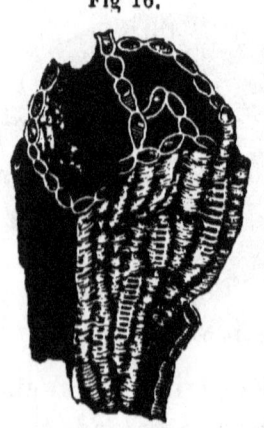
Chain-coral.

THE STORY OF A PIECE OF LIMESTONE. 55

Fig. 17.

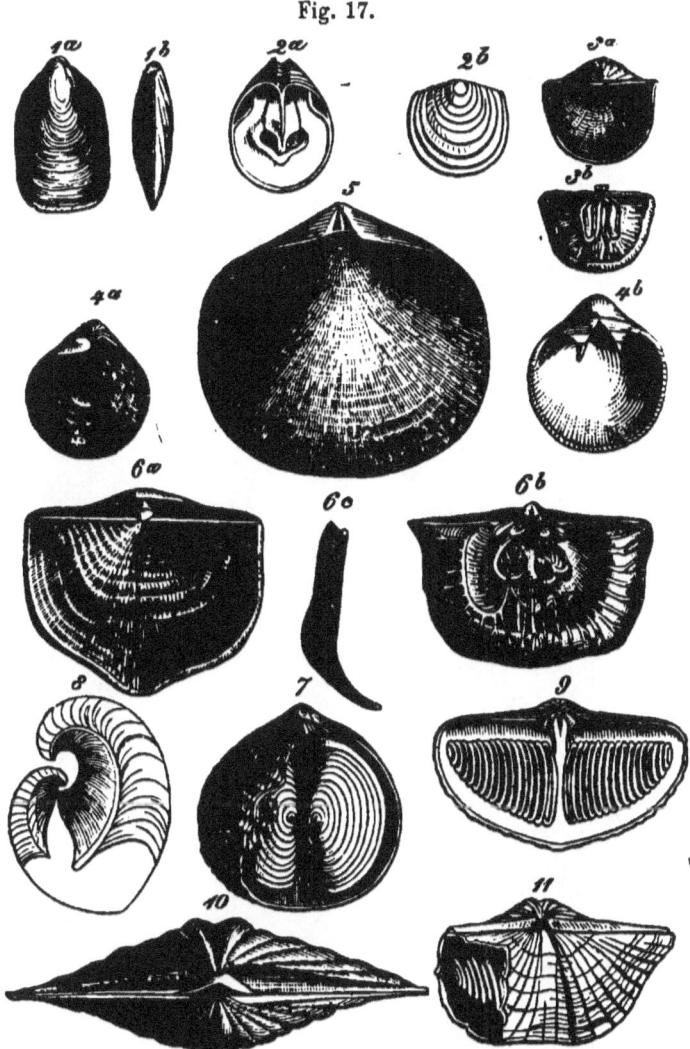

PALÆOZOIC BRACHIOPODS. 1. Lingula Lewisii (Silurian). 2. Obolus Apollinis (Silurian). 3. Leptaena transversalis (Silurian). 4. Orthis elegantula (Silurian). 5. Orthis striatula (Devonian). 6. Strophomena depressa (Silurian). 7. Atrypa reticularis (Silurian). 8. Pentamerus. 9. Spirifer striatus (Carb. Limestone). 10. Spirifer speciosus (Devon). 11. Spirifer trigonalis (Carb. Limestone).

Ludlow bone-bed to which I have already alluded. I am told that off the western coast of Ireland, near Rockall, such a bone-bed is now actually in course of formation; so that if it becomes covered over by succeeding deposits, it may one day present a similar appearance. Of the land plants of the Silurian period I cannot say much; but that the dry land was more or less clad with green I have not the slightest doubt. What makes me feel so confident about this is that the small spores of club-mosses are to be found fossilized in the "bone-bed" I have mentioned. You can only see them with the microscope, but there is no doubt as to what they really are. These spores must have been carried by the land-breezes seawards, and strewn over the surface of the ocean until they sank, and were buried in the deposits accumulating along the bottom, where the bony-scaled and shagreen-skinned little fishes were living, breeding, and dying.

My story is now finished, for the formation of cracks and fissures in our solid rocks belongs to a later time. Of the minerals and metals which were segregated along the walls of these fissures until the latter became "metal lodes," I cannot say; but thus much—that, apart from the numerous fossils contained in us, our rocks will always be esteemed interesting to man, seeing that it is in them that the over-valued metal gold is most abundant.

CHAPTER V.

THE STORY OF A PIECE OF SANDSTONE.

"You may trace him oft
By scars which his activity has left
Besides our roads and path-ways (though, thank heaven,
This covert nook reports not of his hand),
He who with pocket-hammer smites the edge
Of every luckless rock or stone that stands
Before his sight by weather stains disguised,
Or crusted o'er with vegetation thin,
Nature's first growth, detaching by the stroke.
A chip or splinter—to resolve his doubts;
And with that ready answer satisfied,
Doth to the substance give some barbarous name,
Then hurries on; or from the fragment, picks
His specimen."
WORDSWORTH'S *Excursion*.

LIKE my mineralogical acquaintance, the piece of limestone, generally I am about to do duty for a group of individuals common to every geological formation. But each of us has a separate story to tell, and I shall find it quite sufficient to bring all the circumstances of the epoch in which I lived sufficiently clear to my own recollection. It is said that a number of people who live in the present period (so far removed in time from mine) profess to be able to interrogate a piece of limestone or sandstone, by what they term *Psychometry*, and to get its story in some easier way than by the ordinary cross-questioning of science! All I can say is, I wish the events of my own life were so permeated in my substance. If this theory be true, the modern science of geology will have to give up induction, and fling itself into the arms of the spirit-rappers!

Every one of my listeners knows what a piece of sandstone is like. There is no need for me to describe my appearance, therefore, as novelists do their heroes. But how many thus familiar are aware that in ninety-nine cases out of a hundred every such piece of sandstone was originally formed along the floor of ancient oceans? Those ocean bottoms are now represented by dry land surfaces, where the vegetation luxuriates on the mineral substances accumulated under such widely different circumstances. Even where no marine organic remains are present, as fossils, to prove the marine origin of the sandstones, that origin is none the less certain. I cannot speak with certainty as to the nature and extent of the dry lands and continents of the epoch in which I was born. Suffice it to say, they must have been great, for the rivers which watered them were large, and brought great quantities of mud and sand down to the sea. The ocean currents and tides also wore away the coast-line, and added to the quantity of loose sand and mud which accumulated under the waves in consequence. Thus it was that I was born.

My earliest remembrances are of my lying loose and unconsolidated on the ocean-floor, and of constant additions being made to the sheet of which I formed part. It was whilst I was lying in this state, as so much ordinary sand, that I received my impressions of what was going on around me. These consisted of a familiarity with the commoner animals which lived in the sea, or with occasional plants and vegetables which had been carried there by rivers, until they sank to rest in my bosom when they had arrived at a water-logged condition. Of these I will speak presently. Meantime let me make a few remarks as to the changes which transposed me from loose marine sand into hard sandstone; and in doing so, it will be evident that the same explanations will answer for the similar alteration of sandstone rocks, both of earlier and later geological periods.

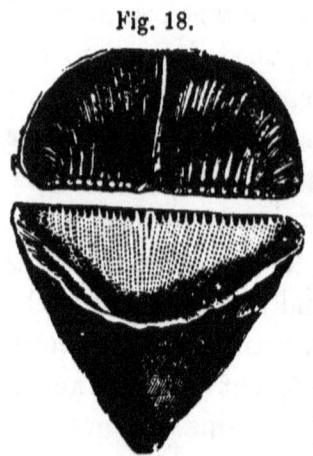

Fig. 18.

Calceola sandalina.

The sand or mud brought down and laid on the sea-floor in the manner I have mentioned was not of an absolutely pure character as regards its mineral composition,—that is to say, it was not all silica, or alumina, as the case might be. In most instances the material was mixed with more or less of iron rust, or of lime, and silica. The two latter acted as cementing pastes to those sandstone rocks which are now of a lightish colour; whilst the iron was the compacting agent with such dark red rocks as that of which I form part. Indeed, in most cases, even when the sandstone is of a light yellow, a small percentage of iron has gone a great way towards binding the loose grains of sand together, and thus producing a hard rock. When this chemical agent has been equally dispersed through the sandy mass, you have the thick-bedded sandstone, or "free stone." When it was intermittent in its action, or unduly mixed up, or occasionally alternated with something else, then the sandstone becomes "flag-stones" of greater or less thickness.

Sometimes you will see a mass of red sandstone more or less mottled. This has been caused, in most instances, by patches of vegetable matter— old world *fucoids* or something of that sort,—which decomposed, and whose chemical changes combined with the iron, and locally prevented its colouring effect.

Of course it will be evident that our hardness or softness greatly depends on the percentage of cementing material, or to the different circumstances under which we were formed. I have no doubt that, when the chemical changes above mentioned were going on through an immense thickness of accumulated sand, the hardening process was

62   THE STORY OF A PIECE OF SANDSTONE.

greatly assisted by the pressure of the overlying volume of sea-water.

The epoch to which I belong is sometimes called the "Old Red Sandstone," and, occasionally, the "Devonian." The former term is given to our formation to distinguish us from the "New Red Sandstone," overlying the coal-measures; whilst the latter name is of local origin, and indicates that the system is largely developed in the lovely county of

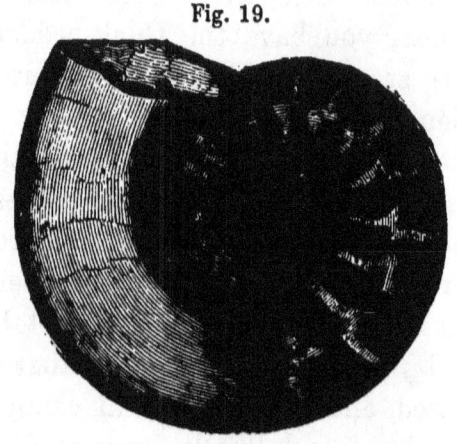

Fig. 19.

Clymenia.

Devon. Indeed, that sunny land owes no little of its physical attractions to the various mineralogical structure of the rocks of our formation. Perhaps I can boast of the fact that there are few other formations which have such a world-wide extent as that to which I belong. In the United States it stretches over an area nearly as large as Europe, there being one continuous coral reef included in it which covers an area of nearly half a million of square miles. In Canada there is also a great extension of this formation; whilst in South Africa its area is greater still. In Russia one sub-division is much greater than the whole of England, and there is a large extension of beds of similar age in Asia Minor, as well as in Australia.

The original name of "Old Red Sandstone,"—given to the formation of which I am a humble part, was conferred upon the thick beds found developed in Herefordshire, Worcestershire, Shropshire, and South Wales, as well as others supposed to be of similar age in Scotland. In the former localities they attain their greatest thickness, which is between eight and ten thousand feet. There geologists have divided the series into four divisions, of which the lowest may be said to blend with the underlying Silurian formation, and the uppermost with the succeeding Carboniferous. In Scotland the beds are not so thick, their greatest vertical accumulation amounting to about four thousand feet. It would seem, therefore, as if the material which formed these rocks came from the south-west, thinning out in a north-easterly direction. In Devonshire, as well as in Ireland, there are two series of strata included in the same formation, which seem to have had quite a different origin. The former indicate a sea in which coral reefs abounded, and the latter tells us plainly of a large continent which existed towards the end of this epoch, on which there were freshwater lakes as extensive as those of North America. Perhaps it was the same continent whose rivers contributed no little of the sand and mud which, when strewn on the sea-bottom, formed the sandstones of which I am part. I am told, however, that there are some geologists who imagine that all these red rocks were of freshwater, and not marine, origin; but I think that their immense area will convince you that this could not be the case.

How shall I tell of the strange sights which I beheld when quietly lying on the ocean-floor! The sea-water had the same specific gravity it has now, and the constitution of the atmosphere was similarly formed. It is an error to suppose, as some have done, that there was mixed a large percentage of carbonic acid in the air before the Carboniferous epoch, and that this was absorbed by the plants, and the atmosphere cleared and rendered fit for animal life at the same time. The theory is ingenious, but there is not the slightest ground for believing it has any foundation in truth. Occasionally the sea-water became turbid and red, owing to larger quantities than usual of the refuse of igneous and metamorphic rocks being carried down by the rivers. As is well known, these contain large quantities of iron, which are easily decomposed, and enter into new combinations as oxides; whence my colour and also my cementing agent. The sea-bottom was covered with groves of *fuci*, or sea-weeds, in which a large crustacean, bearing some resemblance in its huge claws to the modern lobster, lived and left its spawn. The latter is actually found fossilized in our sandstones, and bears some resemblance to a flattened blackberry. Among geologists, I am told,

Fig. 20.

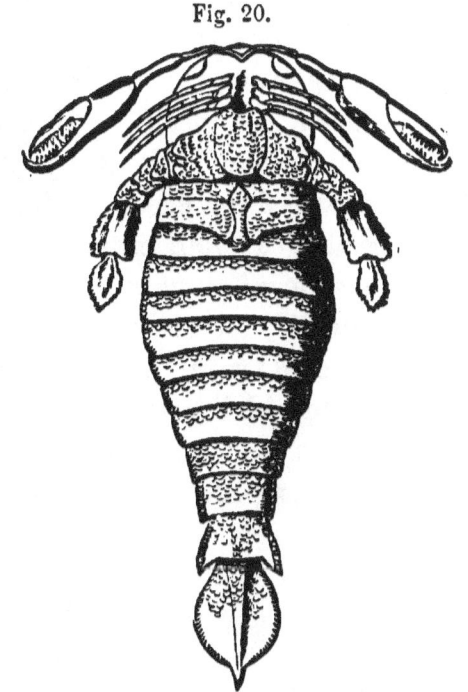

Huge fossil Crustacean (*Pterygotus Anglicus*). Old Red Sandstone, Forfarshire.

it goes by the name of *Parkia*, whilst the huge lobster which left it, and which was at least six or seven feet long, rejoices in the name of *Pterygotus*. Several species of this common form are met with in Scotland, as well as in England. Another large crustacean, which appeared during later Silurian times, and was nearly related to the *Pterygotus*, now goes by the name of *Eurypterus*, on account of the breadth of its swimming feet.

Fig. 21. The Gar-Pike (*Lepidosteus osseus*), 1¼ nat. size, a Ganoid fish, now living in North American rivers.

But by far the commonest creatures which enjoyed life in the sea of my birth were the fishes. Indeed, my epoch has been justly called "the age of fish." In many places they swarmed in shoals. Most of them belonged to an order of which there are very few now living, termed the *Ganoid*, on account of their being covered with a series of oval or rhomboidal bony plates, instead of scales. These bony plates had an exterior varnish; whence their name. At present, I am told, there are several species living in the rivers of North Africa, and others enjoying life in the lakes and rivers of North America. But out of nine thousand species of fish known to naturalists the *Ganoid* species only number about twenty-nine. Indeed, the wide geographical areas where the two outliers of this once numerous and world-wide family of fishes are now lingering, indicate their antiquity, and suggest how many geological phenomena have taken place to bring about their present geographical isolation. By many it is supposed that the whole of this family would now have been extinct, had it not been for their withdrawing from the keen battle of life that subsequently went on in the seas by the introduction of other species, and so confined themselves to fresh-water condition. Few of these peculiar species

Fig. 22.

Pterichthys Milleri.

have a bony skeleton properly hardened, as is the case with ordinary thin-scaled fishes. No doubt the strong, bony integument did duty instead. Indeed, among the fish which lived during my lifetime, scarcely any possessed a solid skeleton. The largest of these strange-looking fish is now called *Asterolepis* from the star-like markings on each of the scales. It reached the entire length of between twenty and thirty feet. Other common forms were the *Holoptychius*, noted for its large oval scales being peculiarly wrinkled; the *Pterichthys*, or "winged fish," so called on account of its two pectoral fins, which are very large and resemble paddles, being placed near the head, where they look like wing appendages. The plates which covered this fish were very large, and ornamented by a series of granules. The former of these two species lived in what is now America, Russia, and England, and Scotland.

Fig. 23.

Pterichthys cornutus.

Then came the *Cephalaspis*, or "buckler-headed" fish, so called because its queer-shaped head was encased in a shiny bony buckler, in form not unlike a cheesemonger's knife. Its trilobed body was covered with lozenge-shaped bony plates. The *Osteolepis*, or "bony-plated" fish, was the most abundant; its name being derived from the minute rhomboidal plates which covered its body, and protected it, like the links of an ancient coat-of-mail. Besides the fishes of this class, which, singularly enough, were further distinguished by their having the tail unequally lobed—and not regularly cleft as in the common herring and other scaled fishes—there were associated with them others, having an affinity with species of the Shark family. These are called *placoid* fishes, on account of the skin being a kind of shagreen, dotted with minute plates or points of hard bony matter. They also have a cartilaginous skeleton, as, for instance, the common skate, sturgeon, &c. Well do I remember the above fish, ranging in size from the *Asterolepis* to the little *Onchus* and *Osteolepis*, of only a few inches in length! The quick, active movements of the latter fishes, as they roamed in and out of the thickets of seaweeds, caused the light to flash from their enamelled scales, and sometimes only too surely pointed out their playgrounds to their cestraciont enemies. They had their feeding and their spawning-grounds, and each of these places is now represented by the greater number of fish found fossilized in the flagstones, as in the Caithness flags, and the yellow sandstones of Dura Den.

Sometimes, also, great numbers were killed by unusual quantities of mud being poured into the water and choking them, as a turbid river will, at the present time, suffocate the smaller of its tribes. How suddenly these died is indicated by the fact that thousands of fossil specimens are to be seen with their fins erect, like those of the *perch* when he is "struck by the angler." Others are contorted and bent, as if in pain; their last dying struggles having thus been faithfully handed down by the stony records in which they were imbedded.

Some few of the fossil fish of this period had *reptilian* characters in their teeth, &c., indicating and linking on, as it were, the next great family which should rule creation. Wherever the Old Red Sandstone has been met with, some, if not all, of these peculiar ganoid fishes have been found fossilized. Therefore they are good indications of the geological age of any such formation.

I will not trouble my listeners with the dry, technical details of how the strata succeed each other in my parent formation. I want, if possible briefly but vigorously to sketch the life-characteristics of that distant epoch.

I have thus far devoted myself to the fossil fishes because of their abundance, and also of their very striking peculiarities. I now come to other creatures, perhaps not less abundant, but not so attractive. I must premise, however, that such marine creatures as corals, mollusca, and trilobites were not very abundant over the area where I first saw the light. They delighted in clearer water, and so are to be found over the area where that existed. Indeed, generally speaking, those parts of the sea-bottom where most of the red muddy matter was poured in were shunned by all forms of life, not excluding the hardier fishes. Hence it is you rarely find, in the very red sandstones any organic remains or fossils beyond a few vegetable impressions.

Of course there were various parts of the same sea thus distinguished by different physical circumstances, and life was developed, or located accordingly. Let me, therefore, give you some slight account of the area where "blue water" was most in force, and where, in consequence, there were the most numerous assemblages of crustacea, shell-fish, and corals.

The localities in Great Britain where these peculiar fossils are found in strata of the age I am describing, lie chiefly in South Devonshire, as well as along the North Devon coast. At the latter place you may see beds of sandstone, red and yellow, alternating with slates, limestone bands, &c., the last-mentioned being especially full of organic remains. The total number of species of fossils of all kinds which have been found in Devon alone is three hundred and eighty-three.

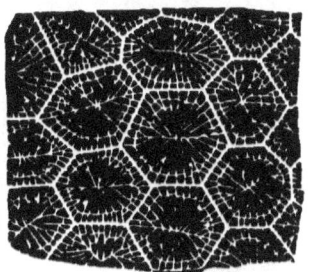

Fig. 24.

Fossil Coral (*Stauria astræformis*).

The highest of the series go by the name of the "Pilton Group," and these are perhaps of the same geological age as the Devonian strata in Ireland. Among the fossil shells which lived during this epoch, and which occur at the above-mentioned places in the fossil state, the most numerous were those belonging to the *Brachiopoda*. Indeed, these shells far out-numbered the ordinary and more highly-organised conchifera, whereas at the present time the latter are by far in the majority.

Among the commonest of the shells I remember were several species of *Spirifer, Stringocephalus,* &c., and also of *Clymenia, Megalodon,* and others. The last was a lamelli-branchiate mollusc, allied to the oyster and mussel of the present day. Among the corals there abounded in Devonshire the *Favosites polymorpha,* or "many-shaped" coral, as well as *Heliolites,* or "sun-coral," *Strombodes,* &c. The latter my readers will readily recognise when I tell them it is the common pink or red variety usually bought at Torquay, and which, when polished in the mass for mantelpieces, has such an attractive appearance. All of them are portions of reef-building corals, and well do I remember the animated appearance of the clear water when the "reefs" flourished in their bright colours, and trilobites, fish, and crustaceans swarmed around the busy pile. The *Trilobites* found in the Devonian limestones are of a peculiar type, equally distinct from those of the preceding Silurian period, or of the succeeding Carboniferous. Among the commonest of the genera were *Brontes,* noted for its fan-like tail, and *Homalonotus,* equally distin-

Fig. 25.

Zaphrentis cornicula.

THE STORY OF A PIECE OF SANDSTONE. 73

guished by the double row of small spines running down the central lobe, and which give to it a more

Fig. 26.

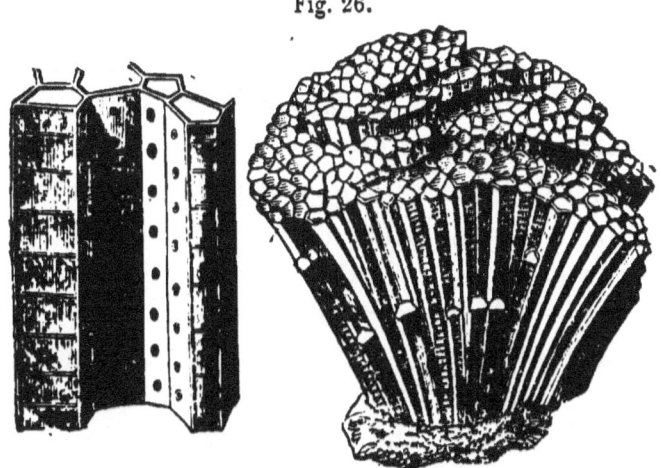

Common Devonian Coral (*Favosites polymorpha*).

Fig. 27.

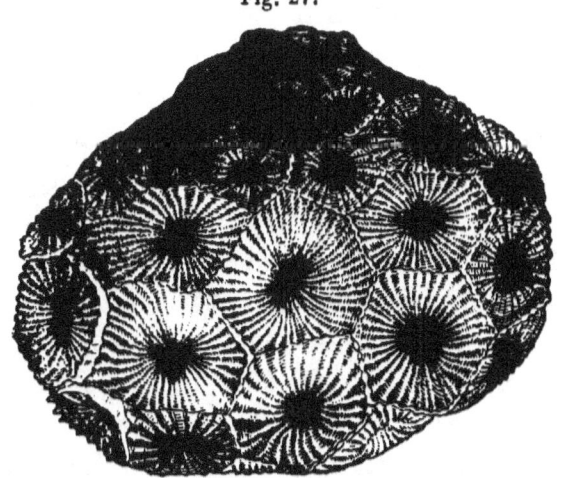

Cyathophyllum hexagonum.

"trilobed" appearance than any other species in the

whole family. But, clear though the sea-water generally was in which these Devonian beds were formed, every now and then shifting currents brought fine mud and other sediments. These were thrown down on the ocean-floor, where they alternated with the bands of limestone.

Eventually, the sea again maintained its purity for a long period, during which the corals and other clear-water-loving animals resumed their avocations, and left behind them traces of their work.

I have said that where Ireland now stands, was part of a great continent, or some other extension of dry land, towards the close of the age in which I was born. Of this I cannot speak with certainty; but the evidence is strongly in favour of the idea. In the country of Kilkenny are a series of fine-grained greenish sandstones, regularly bedded; they are full of evidences of fresh-water deposition. Nowhere, in Europe at least, will you met with such well-preserved land-plants; all of which prove, by the perfect manner in which they have been preserved, that they could not have been drifted from a distance, or been in the water long. Among the most attractive of these remains are those of a tree-fern, formerly called *Cyclopteris*, or "Round-leaved Fern," but now named *Palæopteris Hibernicus*, or the "Primitive Irish Fern." Nothing could be more exquisite than this beautiful fern, even in a fossil state, and you may therefore guess how attractive were its groves when it was the monarch of the primeval forests, and its graceful fronds bent over the clear waters of a lake which equalled in picturesqueness those of the Emerald Island of these times.

This fern is not unlike, in general appearance, the modern "Royal Fern" (*Osmunda regalis*), with the exception that it has no mid-rib—its veins ramifying from the base towards the exterior of the leaf. Associated with this tree-fern were great and small club-mosses, which trailed over the ground, and formed a rich green carpet of various tints. Among the commoner of these extinct club-mosses were *Sagenaria* (of which the seed-vessels and catkins are well preserved); *Psilophyton*, a simpler club-moss, and the larger and more tree-like *Lepidodendron*, which afterwards became so abundant during the Carboniferous epoch.

Besides these we have evidences of other kinds of vegetation, and there is no doubt that the higher grounds were more or less covered with more highly-developed and organized species. What is further corroborative of the fresh-water origin of the Irish sandstones is the immense number of bivalve shells, exactly resembling the large fresh-water mussels (*Anodon*) which abound in modern English rivers. Both in appearance and structure these fossil shells are evidently closely allied, and therefore they are called *Anodonta*. They abound by thousands in some parts of the sandstones, associated with plant-remains, and with those of crustaceans which seem allied to the modern crayfish. So long did these large Irish lakes exist, that mud was strewn along their bottoms which ultimately formed rock several hundred feet in thickness. I am told that similar deposits of fine mud and shell marl are now going on along the floors of the forest-fringed lakes of North America. Change the character of the vegetation there, and you have no indistinct restoration of the Irish Devonian lakes. Many of the fish would do; for the "bony pike," a ganoid fish, still lives there, associated with colonies of "swan mussels" (*Anodon*) clustering on the bottom.

So much for the brief outlines of my story. Much more could be said upon this remarkable epoch; but if I have given anything like an idea of my origin and of the character of the life-forms with which I was brought into contact, my business is done, and I accordingly retire for another geological speaker.

Fig. 28.—Ideal Landscape of the Carboniferous Period.

CHAPTER VI.

THE STORY OF A PIECE OF COAL.

"A passion for plants had so grappled his soul,
That an old *Hortus siccus* each spare moment stole;
For which he had ransacked the swamps and the meads,
Till his *Hortus* was richest in grasses and reeds.
But a strange antiquarian whim he displayed;
From the simplest of plants his selection was made,
And of structure primeval like none we descry
'Mid the bountiful gifts that the seasons supply;
Nor confined he his search,—for the earth widely knew,
From the poles to the tropics the treasures he drew:
Which long in his cabinet hoarded so slily,
As an ancient Herbarium are prized very highly."
King Cole's Levee.

CAN any of my listeners form any idea of what a million of years means? It is very difficult, I grant, but I cannot give any more definite conception of my own great age than by saying I am many millions of years old. Before I attained my *majority* —that is to say, before I became really and positively *coal*— I had existed in manifold forms. You cannot hit upon a greater mistake than to suppose I was originally made just what you now see me—a jetty mass of mineral. The doctrine of metempsychosis, said to be held by the Hindoos, would apply almost literally to my own biography. You may trace my career through a hundred different stages, each more widely various than the other. Nay, the process of elaboration through which I have passed is so complex that I may well be forgiven if I have not a clear recollection of it myself.

I am English born and bred, notwithstanding the seemingly tropical character of my antecedents. In some measure, it may be thought that I hardly partake of English characteristics as regards the climate which affected my earlier career; but I can assure you I was never once removed from British ground. In the distant ages to which I have briefly referred, my recollections go back to waving forests of tree-ferns and gigantic club-mosses, as well as to a thick underwood of strange-looking plants. The name now given to this formation by geologists is termed the Carboniferous, and you may form some idea of the ages which have flowed away since then by the fact that no fewer than *nine* subsequent distinct formations and periods occurred. These are known as the Permian, Triassic, Liassic, Oolitic, Cretaceous (or chalk), Eocene, Miocene, Pliocene, and Pleistocene, to say nothing of the epoch comprehending the human race. To make myself still more clearly understood, it is necessary to state that the formations *newer* than that to which I belong attain a vertical thickness of more than fifty thousand feet! All this mass was slowly formed by gradual deposition along old sea-bottoms, &c., whilst a more than equivalent period of time was taken up in the upheaving and other processes which have elevated these rocks into their present position!

Fig. 29.

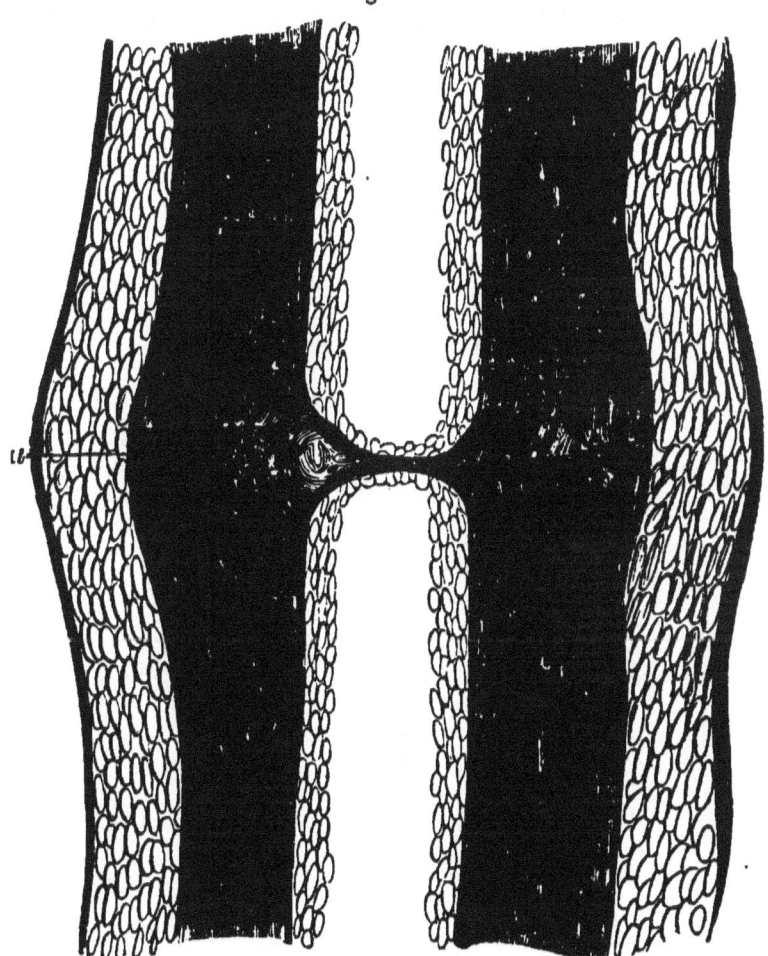

Vertical section of Calamite, cut through node.

The climate and geography of Great Britain were

very different from what they now are when I was born. You must imagine a soft balmy temperature, neither too hot nor too cold, and lacking those extremes which at present characterize the seasons. There was no great necessity for extreme heat—

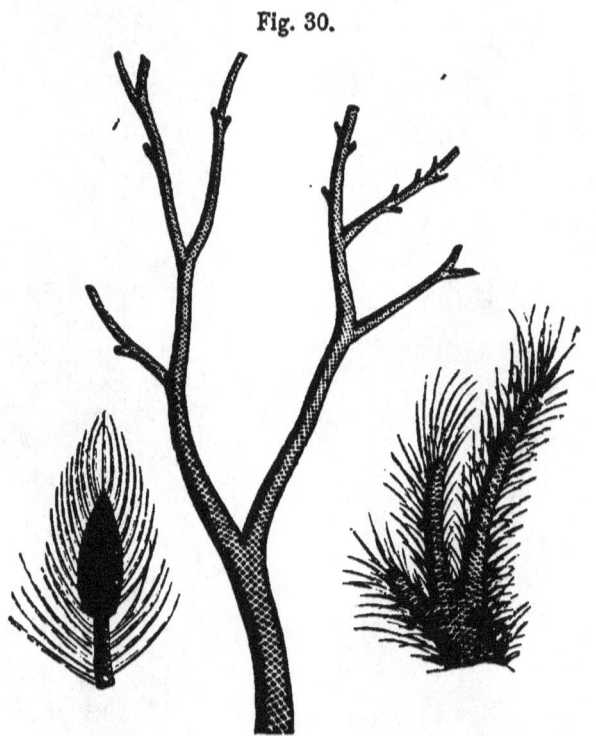

Fig. 30.

Branches, and fruit (*Lepidostrobus*) of Lepidodendron.

rather it was most important to the growth of a luxuriant vegetation to be free from cold. There were few ranges of hills or mountains, for these always cause a refrigeration of the atmosphere by condensing the clouds; thus hanging the sky with

THE STORY OF A PIECE OF COAL. 83

a curtain which shuts off a great deal of solar heat.
True, right across what is now central England,
there stretched a mountainous barrier, perhaps of old

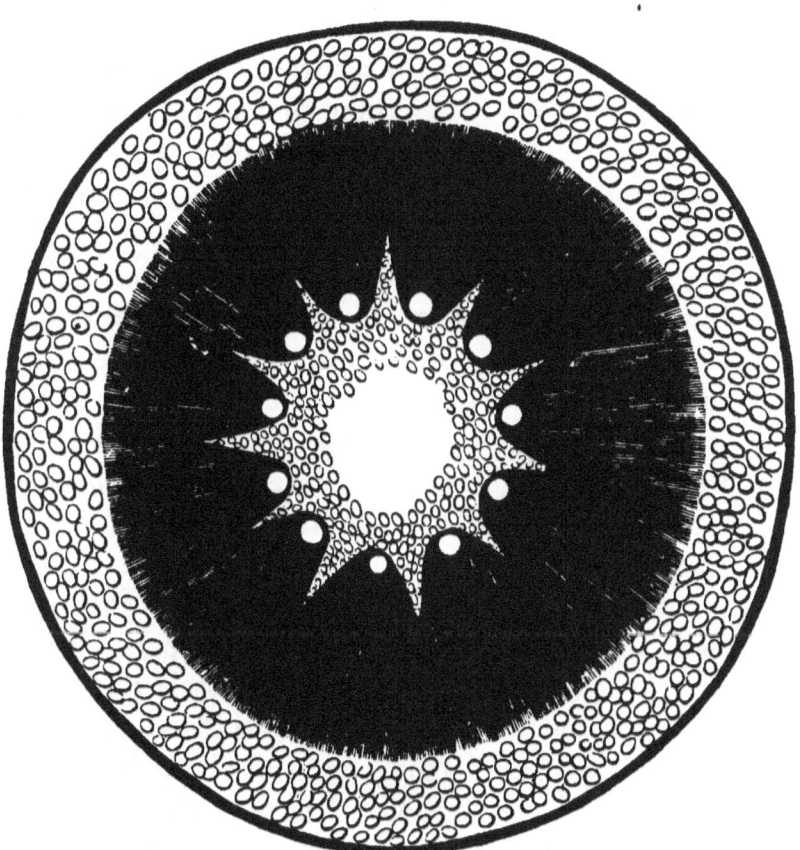

Fig. 31.

Transverse section of Calamite, showing cortical layer surrounding
woody wedges.

Silurian rocks. Scotland and Wales were also
then widely different from what these countries are
at present. Instead of the grand mountainous

scenery they now possess, there were long-extended saline mud-flats, thickly studded with trees now extinct, and known to the geologist by the names of *Sigillariæ, Lepidodendra,* and *Calamites.* In fact, all the district now considered as "coal-yielding" was then similarly circumstanced. The entire area had a geographical condition similar to the marine swamps which now fringe the coast-line of the Southern States of America. To these the slowly ebbing and flowing tides had access nearly twice a day. Around the more aged trunks of these extinct trees, standing on a muddy, shallow sea-bottom, so to speak—marine worms clustered, and their coiled tubes are now occasionally found fossilized, along with the petrified vegetation to which they clung when in life. These *Spirorbi,* as they are commonly termed, are tolerably plentiful in the north of England. It was owing to the semi-marine, semi-terrestrial character of the area on which the luxuriant vegetation of the Carboniferous period grew, that we now find so many fossil mussels and other marine shells imbedded in the same strata.

I am told that chemists nowadays have discovered only one atom or particle of carbon associated with every thousand of the other gases forming the atmosphere. The atmosphere of the period when I was born hardly contained more. This small quantity was absorbed by the waving forests into their structure, and thus added to their solid bulk. Day by day, and year by year, each individual tree grew, so that the mass of solidified carbon increased, but without exhausting the original store. This was constantly being furnished by volcanoes, as well as by the lowly animals of my own time. Everything, they say, is composed of minute and cellular parts, and originally my atoms freely floated in the air as so many particles of carbon. This was before I had entered into that combination which made me part and parcel of a living tree. Once having been sucked into the leaf-pores of a *Lepidodendron* or *Sigillaria*, I started existence under a new form. I became subject to those unknown laws of vital force which philosophers find so great a difficulty in explaining. I had now an active duty to perform, and had to assist in the growth and wellbeing of the tree in whose bulk I lay. But this did not prevent me from noticing the many strange objects which surrounded me. Tree lizards, not very much larger than those which now haunt the sunny banks of old England, climbed up and down the sculptured branches of the forest trees, and lived upon the marsh flies and beetles, whose "drowsy hum" was the only sound that broke upon the stillness of the primeval woods. They found a shelter in

Fig. 32.

Bark of Sigillaria Grœseri.

the hollow trunks of *Sigillariæ*, in association with the pupæ of beetles and other insects. In some places they have been found fossilized together,—a conserved recollection of those bygone times. Great reptiles, resembling frogs, in some respects, belonging to an order called *Labyrinthodonta*, abounded in many parts of Ireland and Scotland. In the former country was a reptile called *Ophiderpeton*, which had a snake-like form, and a compressed tail, so that it very much resembled a water-serpent. No fewer than five different kinds of amphibious reptiles then lived in the very country which now boasts of its freedom from these creatures! It is singular to notice how a great many of the fishes of this period had reptilian characters, whilst the first-introduced reptiles were not only the most lowly organized, but in many respects were related to fishes. Very frequently the salt-water reaches were visited by alligator-like animals, now termed *Archægosaurus*, whose bodies were covered by hard, horny scutes or scales, held together much after the manner a slater now adopts when he tiles a house. These reptiles were five and six feet long, and were adapted to a purely marine life. They were the principal and most powerful animals of the age I am speaking of. In one of the states of North America, Ohio, no fewer than twenty-seven species of reptiles have been found, belonging to ten different genera. Most of them are batrachians, but one has great affinities to the serpents. The atmo-

sphere differed little from its present condition, being neither denser nor more rarified. This you may prove for yourself by the impressions of raindrops preserved in the Carboniferous sandstones. The great drops were driven by the wind aslant, so that there is even now indicated the very quarter from which the wind blew at the time! The passing shower over, the sun peeped forth from behind the dark clouds, and his heat baked the mud, and cracked it, just as he does now the bottom of a clayey pond. These sun-cracks were subsequently filled up, sometimes by sand of a different colour, so that they are fossilized as truly as the shells and plants. The same sandstones yet bear the trail-markings which the marine worms left after they had crawled over them when in a soft state. Occasionally you may even come across their burrows or holes; whilst the flagstones also are impressed with ripple-marks left by the retreating tides!

Although the sea-bottom was so shallow in the neighbourhood of the great forests, I should state that many miles further out it gradually shelved deeper, until there was an area where "blue water" was attained. Previous to the formation of the coal seams, which as a rule belong to the upper part of the formation, the sea was fairly alive with animals of all sorts of natural history orders and classes. Coral banks, with animals putting forth their beautifully coloured tentacles, more various than the rainbow hues, stretched over many leagues of old Devonian rocks, and, as the area was slowly submerging at the time, their united labours, in the course of ages, produced no small portion of what is now termed the "Mountain or Carboniferous Limestone." Shell-fish, allied to the existing nautilus, found in these purer waters, free from land sediment, the essentials of their well-being. In the limestones which their dead shells helped to form there are no fewer than thirty different species of nautilus! They had relatives termed *Goniatites* (long since died out, for they did not possess the hardiness of their congeners), whose chambers were fashioned in a zigzag or angular manner. Then came another group of shell-fish, equally near by blood, the *Gyroceras*, whose coils did not lie so closely together as those of the nautilus. One other class of cephalopods are now known as *Orthoceratites*. They were also chambered, but were straight instead of being coiled. The limestones of this age are crowded with immense numbers both of species and individuals belonging to these genera. Of them all the *Orthoceras* was perhaps the most dreaded, partly on account of its size (some of their shells being three feet long, and as thick as a man's leg), and partly on account of their voracious habits. Fancy them, as I have frequently

Fig. 34.

Goniatites sphericus.

seen them, with their last chamber surrounded with a fringe of long arms, that would indicate no slight danger to bathers nowadays! Hundreds of thousands of these creatures existed. Indeed, they were the scavengers of the Carboniferous seas, eating up everything that came in their way, and perhaps not particular about preying upon a weakly brother when appetite prompted them. In Scotland, in many parts of the limestones formed at this time, the strata, for hundreds of feet in thickness, are

Fig. 35.

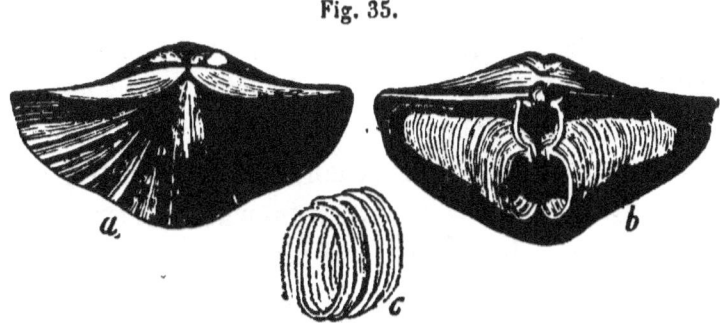

Spirifer striatus, Derbyshire; *b* valve showing internal coil; *c* portion of coil.

composed of hardly anything else but the accumulated shells of *Orthoceratites!*

At the bottom of the sea in which these cephalopods lived and flourished there were gathered together immense shoals of a peculiar shell called *Spirifera*, now extinct. Scores of species of this particular shell lived and died there, for it was the period when the family (*Brachiopoda*) attained its maximum of existence. In fact, they occupied the place in those earlier seas that cockles and mussels do now. Their anatomy was very peculiar, each shell-fish being furnished with a peculiar coiled-up apparatus which it could protrude so as to produce currents that brought to it its food. Small, but beautiful crustaceans, of a race then fast dying out, still swarmed the waters. Formerly you may have heard of them as *Trilobites*—those of this age are christened *Phillipsia*. Their family had exercised a sort of molluscan oligarchy during previous geological epochs. But the Carboniferous period saw the last of the race, and its limestones became their tomb. I am told that the geologist knows few fossils more beautiful than these little trilobites. The cream-coloured matrix in which they are imbedded, and the perfect and ornate characters of the fossils themselves, cause them to be greedily collected and much admired. In the same sea were hundreds of species of shells besides, all of which thronged together to enjoy a common life; but to mention them separately would be to convert my story into a tedious detail. I should be lacking greatly in memory, however, if I were not to mention a most abundant and peculiar family, allied to the star-fishes and sea-urchins of the present day—I mean the *Crinoids*. The common feather-star of recent seas most resembles the upper parts of these extinct animals. But the tentacles of the latter were longer, whilst each was subdivided into branches. When at rest, these closed around the body like the

THE STORY OF A PIECE OF COAL. 93

Fig. 36.

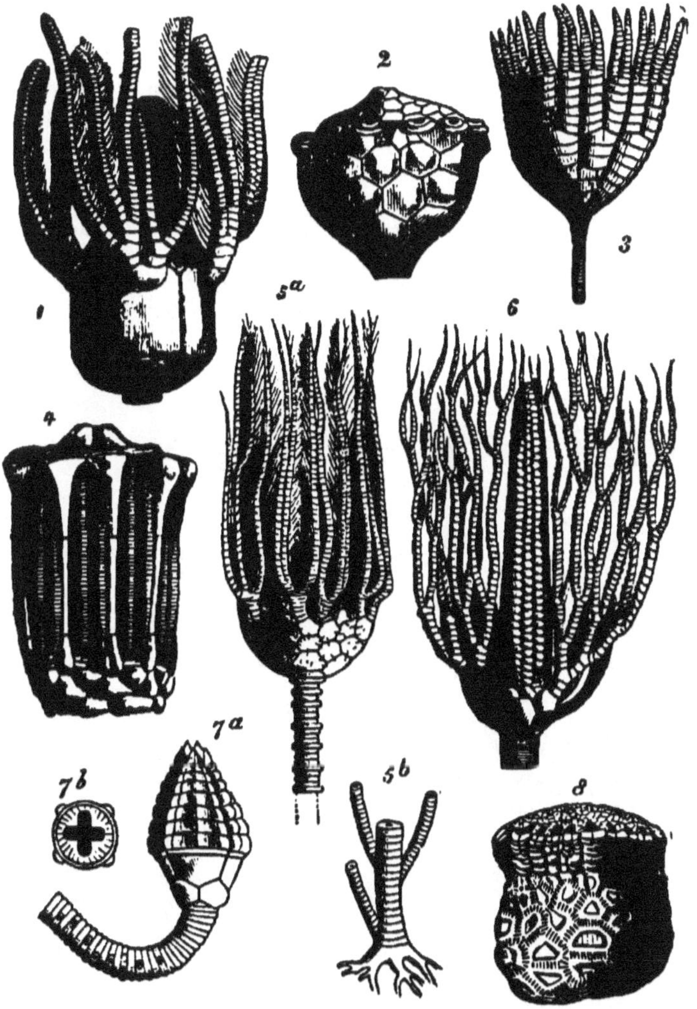

PALÆOZOIC CRINOIDS. 1. *Platycrinus trigintidactylus*, Carb. Limestone, Ireland. 2. *Actinocrinus tricuspidatus*, Carb. Limestone, Belgium. 3. *Ichtyocrinus lævis*, Silurian Limestone, North America. 4. *Eucalyptocrinus rosaceus*, Devonian Limestone, Eifel. 5a. *Actinocrinus triacontadactylus*, Carb. Limestone, Lancashire. 6. *Taxocrinus briareus*, Devonian. 7. *Cupressocrinus*. 8. *Rhodocrinus crenatus*.

petals of a tulip. Again, each was fastened to a jointed stem, which anchored itself by roots to the sea-bottom. Submarine forests of these crinoids covered many square miles of the rockier portions, and their graceful outlines and motions in the water, as well as their bright colours, were sufficient to induce admiration. In Derbyshire the limestone is almost entirely composed of their broken and aggregated stems, compacted so firmly as to form a marble capable of receiving a high polish. I have no doubt you may have seen mantel-pieces formed of it, and have wondered at the strange forms which seem to be enclosed in the solid rock.

As these dead shells and other animal remains accumulated along the ocean-floor to form a limestone that should afterwards be easily identified by their imbedded forms, almost every individual was coated by minute sea-mats. No Honiton lace of the present day ever excelled in grace and elegance that which belonged to these lowly animated beings. In the solid masses of the Carboniferous limestone you may find them festooning shells and corals; and few objects afford greater delight to the geologist when he comes across them. The *single* corals also—that is

Fig. 37.

Pentremites florealis, Carb. Limestone, Illinois, United States. This fossil almost makes up the bulk of the limestone.

to say, those which did not grow in reefs, but lived solitary on the sea-bottom—were not inferior in beauty to any now existing. Their fringe of gorgeously coloured tentacles made them appear like so many animated flowers; and thus the dark caves of ocean then bore many a flower that was born to blush unseen. Slowly, through countless myriads of years, the Carboniferous limestone increased to its present thickness, principally by the

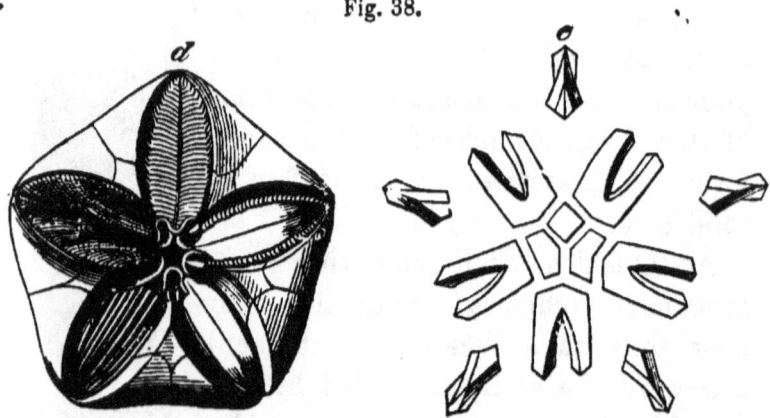

Fig. 38.

d *Pentremites*, enlarged. e Showing the plates, &c., which make up the test.

accumulation of dead shells. The sea-water contained more or less of carbonate of lime, which the shell-fish absorbed in order to build their dwellings, just as the trees did carbon that they might form wood. In this way the minute particles became ultimately condensed into rock masses. Meantime, the water was animated by little creatures that would have evaded human eyesight, although their forms were not a whit less elegant and graceful than those of their larger neighbours. Their tiny shells fell to the sea-bottom, and there formed a limy mud, which acted as a fine cement for the bigger fossils. As time passed on, the sea actually became shallower, by reason of the vast numbers of organisms lying on its floor. The weight of sea-water pressed them into a solid limestone rock, such as you now behold it. Can you wonder, after this, that such a deposit should take a high polish when worked, or that the marble thus produced should be speckled and marked by so many strange forms as you see it in your mantel-pieces or pillars?

In the shallower waters of the sea, and sometimes even in the marine lagoons where the trees grew, multitudes of strangely-clad fishes swarmed. The largest of these, the *Megalichthys*, or "great fish," possessed characters which linked it to the reptile family. Its teeth and jaws rendered it a formidable assailant, and its powerful build and rapidity in swimming made it the terror of its neighbours. In fact, the "great fish" occupied a place among the fishes of its time similar to that held in modern rivers by the pike; its size, also, being about the same. Time, however, would fail me to enumerate the various kinds of fish that lived in the same epoch that I did. From four or five feet in length, to thousands no bigger than the common stickleback, nearly all were covered with enamel plates instead of horny scales. Indeed, horny-scaled fishes did not

H

come into existence for ages afterwards. In many parts of Lancashire, in the shales which overlie the coal-seams, these shining enamelled plates may be turned up by the thousand. The smaller fishes haunted the shallower lagoons overhung by club-mosses and ferns, and the dim light that broke

Fig. 39.

Teeth and Scales of Carboniferous Fish.

through these was often reflected from the sheeny mail of *Palæonisci*, as they wantoned and gambolled, unaware of "great-fish" lying near. When the muddy bottoms of these reaches and lagoons became afterwards hardened into coal-shale, the dead fishes lying there, whose hard covering had protected them from decay, were entombed and passed into a fossil state.

Fig 40.

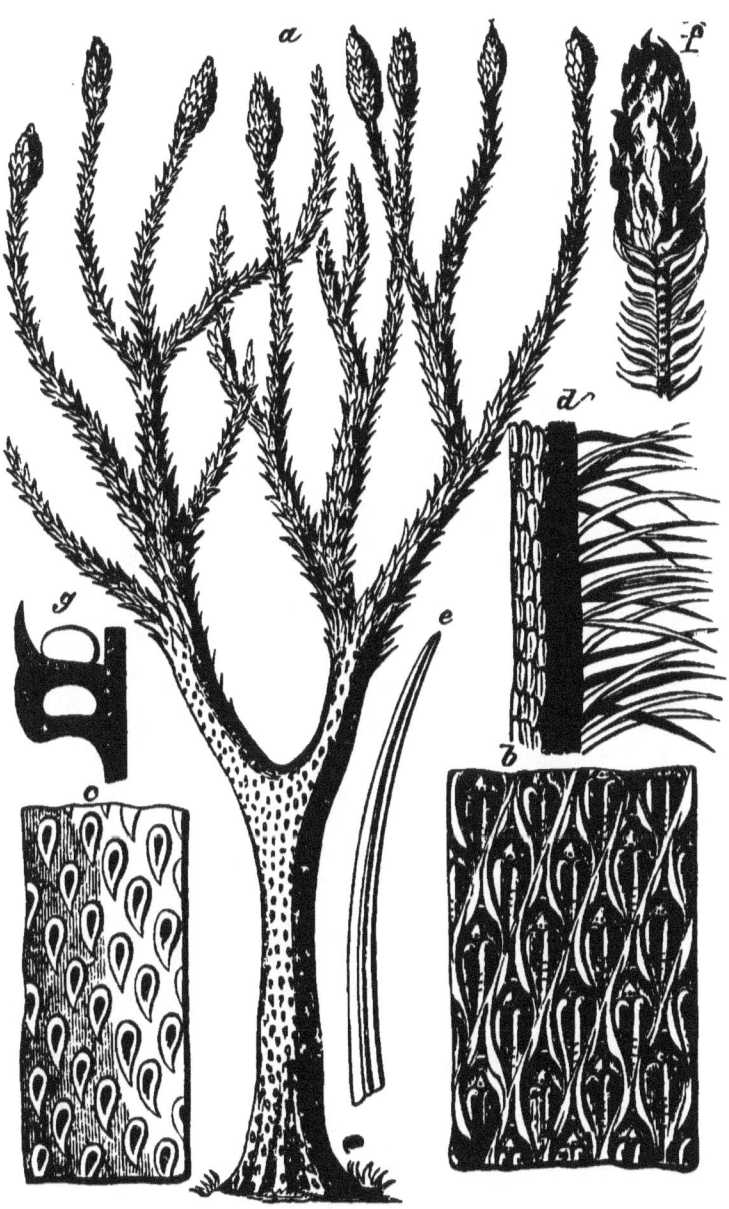

a Lepidodendron (restored); *b & c* impressions on back; *d* stem with leaves; *e* leaflet; *f* fruit of Lepidodendron, called *Lepidostrobus*; *g* showing spores in bracts of fruit.

Fig. 41.

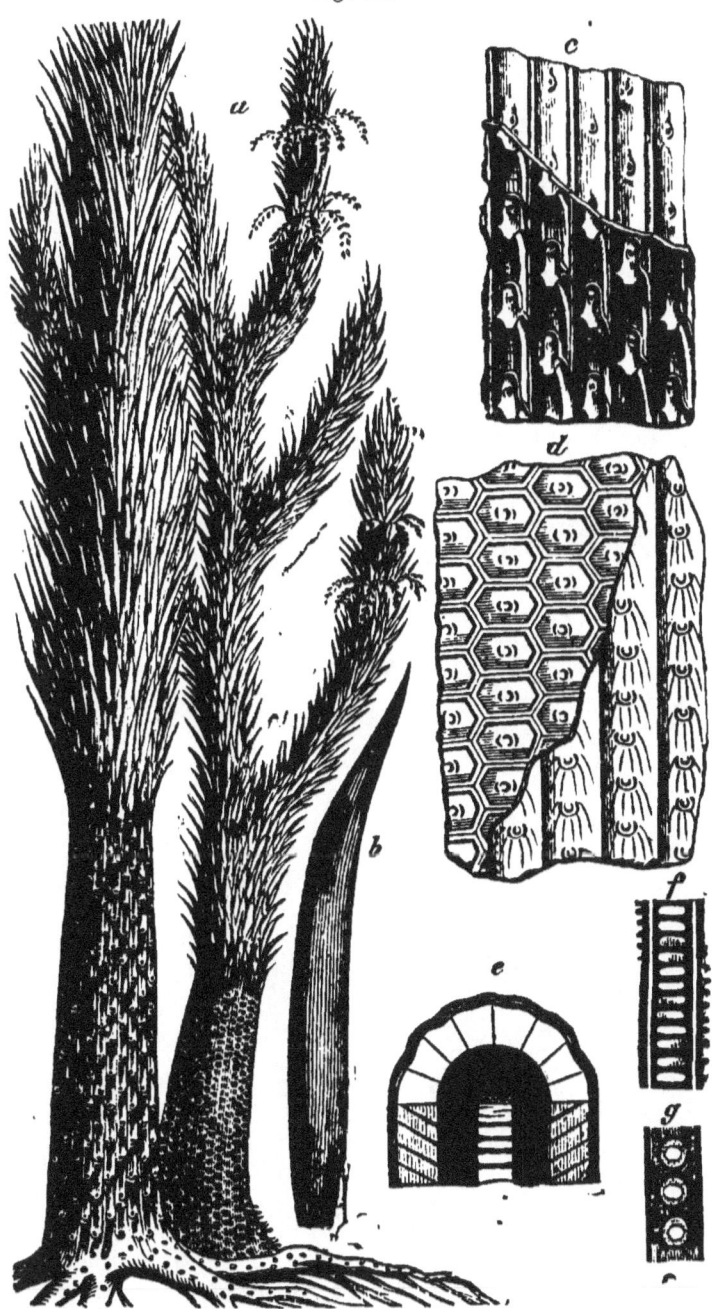

a Sigillaria (restored); *b* leaflet; *c & d* impressions on bark; *e* section of stem; *f* portion of cylinder, magnified.

But what tongue can describe the vegetable wonders of the forests where I grew? The woods were so thick, and the gloom so impenetrable in consequence, that it required a keen eye to make out individual peculiarities. Fancy *Lepidodendra* four or five feet in diameter, and as much as fifty or sixty feet high, and yet nothing but gigantic "club-mosses!" Their long leafy ribbons waved like the leaves of the aspen, and, where these had fallen off, the bark was most gracefully and geometrically patterned from their attachment. Thirty or forty different sorts of these immense club-mosses existed at the same time, each characterized by different leaves and bark. The gigantic *Sigillariæ* were nearly related to them, the main difference being their longer leaves, straighter stems, and the larger marks made on the bark. The roots, also, of this latter class of trees were very peculiar, and stretched through the mud on every side, seeking a firm foundation for the tree to which they belonged. Shooting many feet above these great club-mosses were huge "horse-tails," as easily distinguished from the rest as

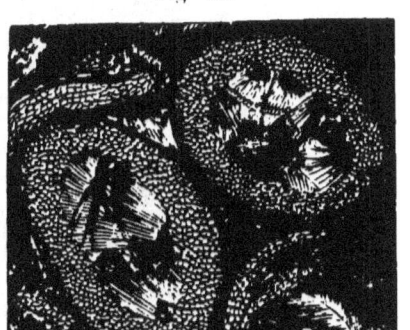

Fig 42.

Microscopical section of Fossil Wood, from clay iron-stone nodules; Oldham.

the aspen-poplar nowadays is from oak and elm. These are called *Calamites*, and truly they were extraordinary objects. You have only to magnify the little

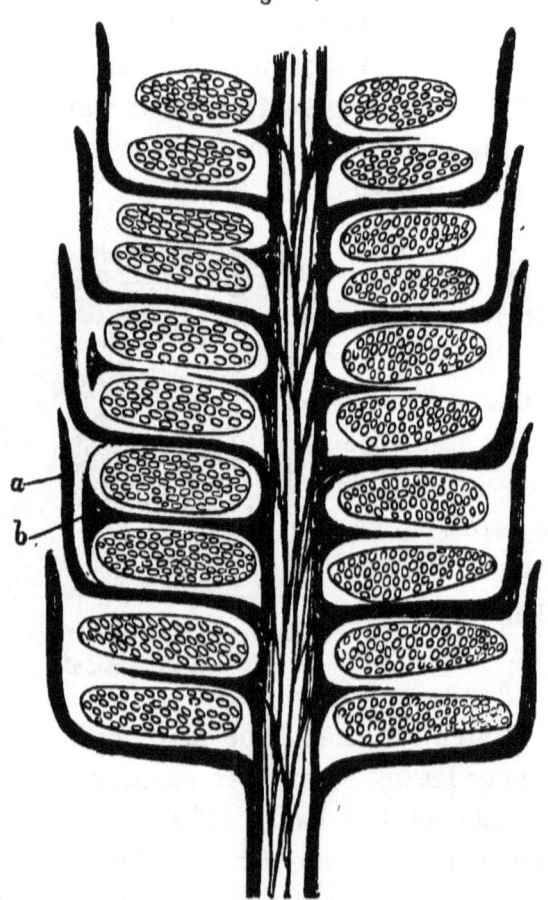

Fig. 43.

Vertical section of Fruit of Calamite, magnified.

"horse-tails" now growing in ditches, until you see them fifty and sixty (or more) feet high, and you would have the best restoration of these Calamites

THE STORY OF A PIECE OF COAL. 105

that could be imagined. There were many species, characterized by fluted joints, and by difference of foliage. Here and there, but more sparsely scattered, were graceful tree-ferns, whose former fronds had left great scars on each side the trunk. The higher grounds were occupied by peculiar species of pine, bearing great berries as big as crab-apples. The

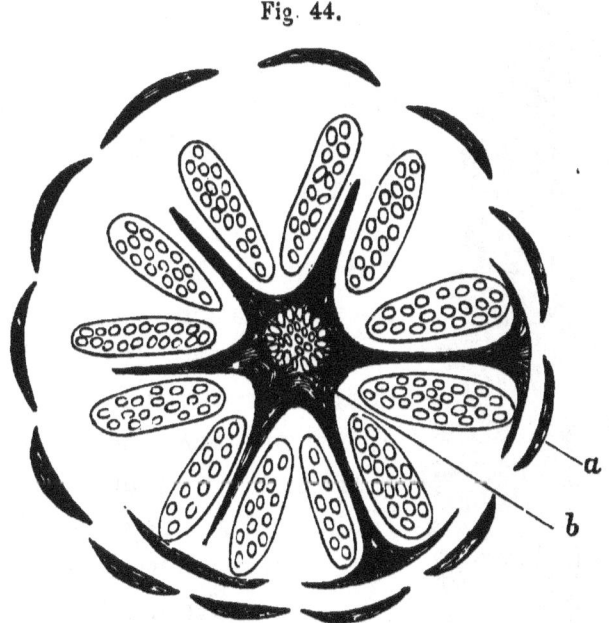

Fig. 44.

Transverse section of Fruit of Calamite, magnified.

humid morass was densely covered by a thick underwood of smaller ferns, which grew there in rank abundance. The equable temperature, rich soil, and humid atmosphere were just the needful accessories to the growth of vegetation of the class I have mentioned. It consequently flourished at a rate of which we can form but a poor idea from the present. The accumulated trees, ferns, &c., were very great, and these gathered in immense quantities over the entire area. I mentioned before that there was a slow sinking or submergence going on. Well, occasionally, the tides brought up silt and strewed it over the decomposing vegetation. In fact, many of the forests were actually buried thus, and their trunks are frequently met with standing erect in solid sandstone rock. But though the covering-up of the vegetation prevented the liberated gases from escaping, it also obstructed for a time the growth of other trees. The latter could not well flourish on sandbanks, and so they were limited to conditions elsewhere similar to those I have mentioned. But as time elapsed, the old circumstances returned. Another forest grew on the site of the older, to be buried up in its turn. During countless ages this alternate growth and covering-up went on, until in some places, as in the South Wales coal-field, there are no fewer than one hundred

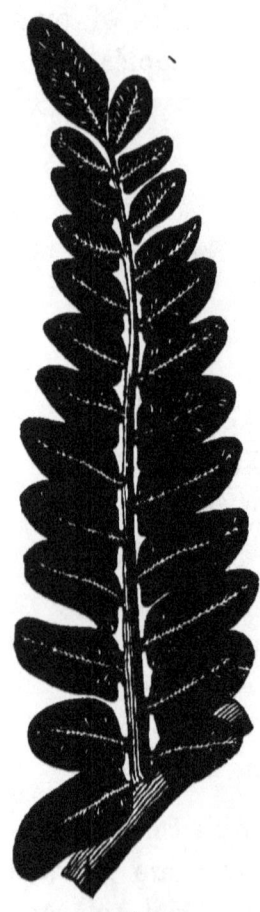

Fossil Fern
(*Neuropteris*).

Fig. 46.

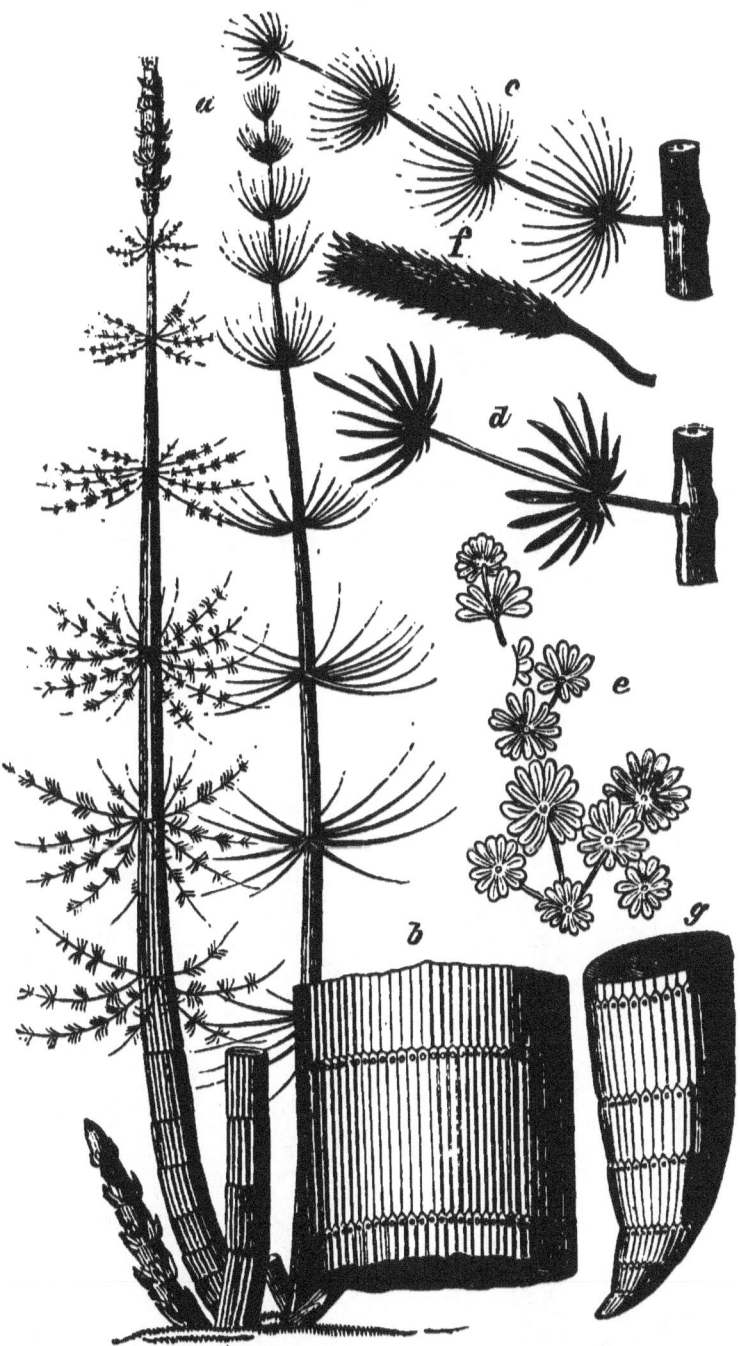

Calamites (*a* restored); *b* enlarged fig.; *c, d, e,* leaflets and branches;

Fig. 47.

Portion of Fossil Tree Fern (*Pecopteris arborescens*) on Coal Shale.

different seams of coal, under each of which you may see a clay full of the roots and rootlets of the vegetation I have been mentioning.

Fig. 48.

Annularia. A Fossil Plant allied to the Calamites.

After this vegetation had been thus collected,

chemical changes began to take place. The mass heated and turned black, just as a stack of hay does now when it has been packed in a damp state. By-and-by, it was transmuted into a pulpy condition, wherein almost all traces of vegetable structure became lost. It afterwards changed into a solid subcrystalline mass, and obtained the jetty, semi-cubical character it now presents. As many of the tissues of coniferous trees contain more or less of silex, which is indestructible, it follows that when coal is burned, this drops out of the grate as a white ash. When the microscope is applied to it, the peculiar dotted vessels of these ancient trees are plainly visible. But notice the associations which cling to a piece of coal! It represents a more solid condition of carbon than is to be found in mere wood. And here I should state, that though various conditions of fossil fuel are met with, from peat wood to culm and anthracite, their vegetable origin is never once lost sight of; whilst chemistry steps in with an easy statement of how these changes occurred! The ancient vegetation of the Coal period grew by virtue of the stimulus of the sunlight. The heat and light induced growth, and thus even a piece of coal represents so much fossil sunshine! So that, when men light their fires or manufacture their gas, they are but setting free the light and heat of the sun which poured down on the old Carboniferous forest, and were stored up by the vegetation in their tissues. Nay, more, botanists will tell you that the three primary colours of light are sure to be developed at some time or another in the history of every plant or tree—in the blue and yellow which form the green of the leaves, and in the red of the fruit or russet of the bark. Just so with the fossil vegetation termed coal. The very aniline colours obtained from coal tar are the restoration of the primary colours which the ancient vegetation stored up from the light! Such is a portion of my history, briefly sketched; but the broad traces of design manifested in my preparation are too palpable to be overlooked. In my mass is stored up a force that saves the wear and tear of human muscle and sinew, that does away with the fearful toil which makes simple slaves of men, and enables them to gain daily bread by easier means. Through the vast ages during which I have been silently stowed away, plutonic disturbances have repeatedly broken through and cracked the solid strata as "faults," and have thus brought them to the surface to enable men to work the coal they contain. Had it not been for this series of subsequent disturbances and breakings, the huge coal-rocks would soon have dipped away beyond human reach, and their valuable treasures have been lost to the world. It is those very agencies, therefore, which men in their ignorance have regarded as sure proofs of Divine anger, that have prevented such a misfortune, and have been among the greatest blessings that have occurred!

114 THE STORY OF A PIECE OF ROCK-SALT.

Fig. 49.—Ideal Landscape of the Triassic Period.

CHAPTER VII.

THE STORY OF A PIECE OF ROCK-SALT.

> "Lives of great men all remind us
> We can make our lives sublime,
> And, departing, leave behind us
> *Foot-prints* on the sands of time."
> LONGFELLOW.

IN many respects I differ from my geological associates, although my story, like theirs, will help to fill up the great lapse of time demanded by the antiquity of the globe. My origin was perfectly natural, and not of that

THE STORY OF A PIECE OF ROCK-SALT.

semi-miraculous nature which some people have imagined. But truth is stranger than fiction, as my own case well exemplifies. I ought in justice to mention that, in the interval between the Coal period and that when I was formed, there was a sort of connecting epoch known as the *Permian*. Geologically speaking, it was not of very long duration. The " Magnesian Limestone " of Nottingham and Durham, &c., is included in it; and its chief and most interesting characters are the probable evidences it affords of a *cold* climate, when icebergs and glaciers existed, and formed what are known to geologists as the " Permian Breccias." With this exception, the general fauna of certainly the greater part of the era greatly resembled that of the Carboniferous period.

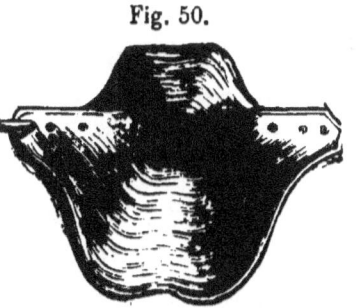

Fig. 50.

Productus horridus, a characteristic Permian fossil.

As a mineral I may lay claim to be almost as well known as my neighbours the pieces of coal and chalk. Geologically speaking, I am not limited to any particular formation or epoch, although I am about to speak of my experiences of that period which has been called "saliferous," or " salt-bearing," on account of the larger quantities of rock-salt to be obtained from it. But in almost the same mineral form I am found in other deposits, from the Silurian up to the Tertiary. In England, however, it is in that formation known as the "New Red Sandstone," or "Trias," that I occur most considerably. In Cheshire my presence is indicated by natural brine-springs, by the disfigured surface of the earth near the salt-mines, and by the dark, thick clouds of smoke from the salt-works which stretch across the heavens.

But before I proceed to describe, as well as I am able, the agencies which were at work elaborating me into the natural condition in which I am now found, or to give you my faint recollections of the physical geography of the period, and the animals and plants which lived—let me borrow a few general remarks from books, as to the classification of those rocks to which I here belong. Their modern name of "Trias" is derived from the tripartite division into which they are separable. These go by the name of "Bunter," "Muschelkalk" (a German name for "shelly limestone"), and the "Keuper" beds. The former prevail largely in Lancashire, Cheshire, Shropshire, Warwickshire, &c., and are noted for their deep red colour, as well as for their thick beds of hardened gravels, or conglomerates of liver-coloured quartz. These indicate rough action in the seas where they were deposited, and the much-worn, rounded pebbles tell an equally plain story of the wear-and-tear to which they have been subjected since they existed as angular fragments of Old Red Sandstone and other rocks.

But throughout the whole of this series, you look

almost in vain for any fossils. The coarse conditions under which the beds were formed were antagonistic to the preservation of any organic remains.

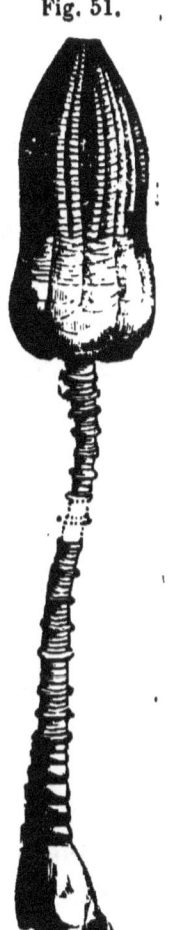

Fig. 51.

Towards the conclusion of this period, in Germany there existed a tolerably deep sea. The waters were pure and free from mechanical sediment; and here the corals and encrinites found all the fitting circumstances for their luxuriant growth and procreation. The sea-bottom was alive with the latter; one particular form, whose elegance has given to it the name of the "Lily encrinite," being peculiar to this particular member of the rock series. The coral reefs increased in the shallower places, whilst amid all these swam great fishes, whose teeth proclaimed their marked *reptilian* affinites, or still huger marine reptiles. Some of the latter had their teeth especially formed for crushing the shellfish on which they fed, and which swarmed along the sea-bottom in countless thousands. Among the latter you detect forms which belong to the Palæozoic as well as to the Mesozoic epoch—forms which geologists not long ago imagined were limited entirely and separately

Lily Encrinite (*Encrinites moniliformis*).

to one or the other of these two great divisions of time.*

It is true the bed containing this admixture of Old World forms is slightly *younger* than those I am more particularly dwelling upon. But I could not forbear drawing the attention of my readers to this striking fact—that the so-called "breaks" in the continuity of organic remains are fast disappearing before a more general geological investigation. The

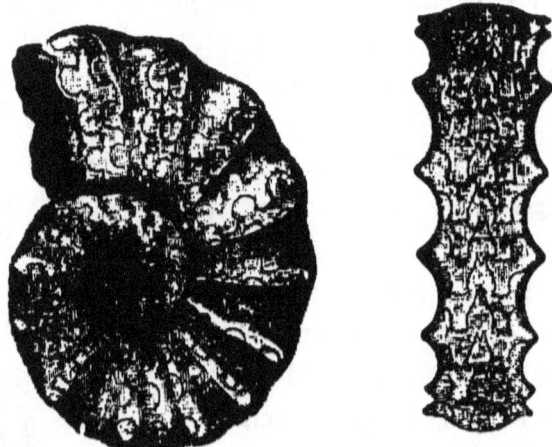

Ceratites nodosus, Muschelkalk; a characteristic fossil.

Hallstadt and St. Cassian beds, occupying the bases of the Austrian Alps, were formed along a sea-bottom during later Triassic times, where the fauna

* The Hallstadt and St. Cassian beds, which belong to the Upper Trias (Keuper beds), are remarkable for containing fossils, such as *Goniatites* and *Orthoceratites*, which are undoubtedly Palæozoic forms, associated with *Ammonites* and *Belemnites*, which are equally peculiar to the Secondary rocks.

of the old and newer worlds met and commingled as on a common platform.

But it is to the third division of this interesting formation that I must specially allude. The middle member, the "Muschelkalk," is absent in England, so that the Keuper beds are seen in many places in midland and northern England reposing directly upon the Bunter. Where this occurs there is usually an "unconformability" between the two. That is to say, the dip of the two sets of strata is different. This means that the lower had been elevated before the upper had been deposited, and therefore it indicates a *break* in time between the two, and shows us plainly they were not continuously deposited.

The Keuper beds are my home. Here was I bred and born! From the top to the bottom, you have ample evidence of the physical circumstances under which they were deposited. Every layer indicates *shallow water;* in the ripple-marks, sun-cracks, rain-drop pittings, and feet-impressions of extinct reptiles. In Cheshire this series contains beds of rock-salt and gypsum, the whole attaining a thickness of fifteen hundred feet. The beds of rock-salt of which I am a humble portion, frequently attain the thickness of a hundred feet; and the area, in Cheshire and elsewhere, over which these extend, is calculated to be above one hundred and fifty miles across! This represents the magnitude of the natural salt-pan where I was formed. The beds are usually split up by a layer of clay or marl, and the rock-salt masses are usually tinted with a dirty red, caused by the slight admixture of iron. But not a trace of a fossil or any other organic remain do you ever get in the neighbourhood of the salt-bearing beds! Farther away, on what would be the flat shores of the sea where the salt was precipitated, you get evidences of fish and reptile life; as in Shropshire, Cheshire, Leicestershire, Warwickshire, &c. Mechanical impressions, such as ripple-marks and sun-cracks, are plentiful enough in the true salt-bearing series; but no *vital* evidences!

What does this general absence of fossils mean? It is not that they could not be preserved, for you have seen that other impressions are well enough, and accurately enough, laid by. It must mean that, in such limited areas at least, life from some cause or another was excluded. Such was actually the case. The shallow sea was so salt that no animal life could exist therein. You have similar conditions now in existence. The Dead Sea, extensive though it is, has no fauna. Its waters are thoroughly desolate, and know nothing of the pleasures of life. They are nothing but a vast menstruum, in which chemical solutions are so thick, that precipitations of the surcharges are constantly occurring. The Dead Sea level is nearly a quarter of a mile below that of the Mediterranean, and I am told that the neighbourhood is marked by Dead Sea beaches, indicating that the waters have been shrinking for generations bygone. The river Jordan continues to pour in its waters, which waters are more or less charged with mineral matter held in solution. The Jordan waters, however, are all evaporated from the Dead Sea surface, and, as the mineral matter cannot be disposed of in the same way, there is no alternative except precipitation. This is actually going on, and I am told that solid, cubic crystals of pure salt may be dredged from the mud of the Dead Sea bottom.

As well as I can remember, the physical conditions of the Keuper sea—at least over part of the Cheshire area—very much resembled those now in action in Palestine. The shells and thin flagstones of the Keuper elsewhere are frequently marked by the cubic pseudomorphs of salt, indicating that, far away from where the salt was most rapidily forming, the water was supersaturated. The absence of molluscan and fish life in the Dead Sea will enable you to understand the reason why the Cheshire salt-bearing beds contain no fossils, although they are so thickly crowded with evidences of ordinary atmospherical and mechanical action. When these beds were deposited, a "Dead Sea" existed in Cheshire and Worcestershire, and for so long a period that these thick, massive beds of rock-salt were formed along its bottom by the simple action of precipitation. We may regard these massive beds, therefore, as locally representing the excess of salt—just as ironstone bands represent the excess of iron, and coal-seams the excess of carbon. The only difficulty which appears is the comparative purity of the rock-salt layers, and this the element of time sufficiently explains. It is very evident that the physical conditions remained unchanged for a long time, otherwise the rock-salt would have been intercalated with layers of other material. The stratum of shale or marl which separates the two main beds indicates a temporary suspension of these circumstances, after which the older conditions returned and lasted until an entire change had set in. The salt-masses are more or less rudely crystallized into columns, but I believe this was a subsequent process to the formation of the salt itself. Of course the brine-springs, from which so much of the salt of commerce is now extracted, have been formed simply by the surface water percolating the beds, and dissolving some of the solid salt in its course. At its exit, at a distance from the rock-salt masses, it is then charged with this culinary mineral. In many parts of Cheshire the surface is dotted with "meres," or fresh-water lakes, the haunts of rare birds and plants, and the prettiest spots to be found in Old England. In many cases—perhaps in all—I believe these to have been formed by the slow settling of the over-lying rock-masses over the hollows left by the dissolving of the rock-salt beneath, in the way I have mentioned. I am told that in coal districts it is very common for the upper rocks to settle over the emptied seams, and to leave hollows on the surface.*

* Some of the canals running through the coal-fields have been

I have simply given you my own idea, to the best of my recollection, of how rock-salt was formed. I have heard others repeat their own, and if you like I will give it you, so that you may take them all for what they are worth; they have supposed a portion of the sea to be separated from the rest by a bar of sand, over which the ocean-waves every now and then toppled to supply it with water. In these cut-off seas or lakes, evaporation was going on, and a corresponding precipitation of salt; the toppling water of course supplying the place of that which had been evaporated. It is certain that rock-salt contains many of the same minerals as those usually met with in sea-water; such as iodine, bromine, magnesia, &c. So far, therefore, the argument is in favour of a truly marine origin of salt. And the occurrence of fish, reptiles, mollusca, &c., in beds of about the same age as those of central Cheshire, indicates the extension of a sea in which the water was fitted for animal life. However, in either of these opinions, the same principle lies at the bottom; viz., that rock-salt was precipitated from the surcharged saline water, and that evaporation by solar heat was the immediate cause!

And now allow me to give you an idea of the animals which lived on the dry-land surface at the time when these important economical stores were

---

so influenced by the depression of the area that their banks have had to be raised to keep in the water. Hence, in many places, they are more than a dozen feet in depth.

being laid up. First, there were several species of a great *frog*-like reptile, or Batrachian. This type had come into existence during the Carboniferous epoch, although such primeval forms seem first to have been purely marine in their habits. During the Triassic epoch, however, they certainly existed as land reptiles. The largest of these great frogs was about the size of a small ox; their teeth are of a very peculiar labyrinthine structure, and this character is very persistent. Singularly enough, the feet-impressions of these reptiles were found by geologists long before any of their remains had been met with. Owing to their remarkable likeness to an impression left by the human hand, the hypothetical animal leaving them was named *Cheirotherium,* or the " Beast with the hand." Another reptile, which combined lower with higher reptilian characters in a very extraordinary manner, was the *Rhynchosaurus,* or "Beaked Saurian." It had the features of a turtle, as regarded its horny bill, combined with the characters of a true lizard. It seems to have been web-footed, for in many parts of Shropshire and Cheshire the sandstone flags are marked as thickly with its webbed feet-marks, as the margin of a clayey pond is with those of ducks! This reptile was not so large as the first I mentioned. The *Labyrinthodon,* as that is now called, seems to have haunted the shores of the Keuper seas and lakes, for its foot-marks are found at many levels. They are generally seen traversing ripple-marks, sun-cracks, &c., as though the creature had passed over between tides.

In America, the same geological formation is impressed for more than a thousand feet in thickness, with the crowded foot-prints of supposed extinct *birds*. Everywhere you have evidence of slow subsidence—a subsidence that was first compensated for by the amount of material deposited over the sub-

Fig. 53.

Labyrinthodon (restored).

siding area. You may often trace for yourselves something of the habits of the above-mentioned singular and extinct British reptiles, so well have the soft sandstones done their duty in recording what they felt and saw! Here the Labyrinthodons slowly lifted their feet from the soft mud, from which there dropped portions before they were next set down. Or you may trace where they sluggishly squatted down, or where their huge bellies trailed over the soft ooze!

But by far the most interesting of the inhabitants of the dry land were small warm-blooded animals, belonging to the lowest division of the class—the *Marsupials*, or "pouched animals." These are now inhabitants of Australia, and Tasmania, and North America—their isolated distribution proving their vast antiquity. In the times intervening since they first made their appearance, species belonging to this group have lived in various parts of the world. That to which I am alluding is very remarkable, as being probably the first warm-blooded mammal which appeared on the earth! Its name is *Microlestes*, or the "little thief," so called on account of its insectivorous habits, as indicated by its teeth. This little creature—for it was not much bigger than a rat—preyed on the insects which then abounded in the pine-forests, or amid the thickets of fern and club-moss. Its predaceous character, however, leads some geologists to infer that it was not the first of its class, but that probably an earlier and simpler type appeared before it.

In a bed of later date, formed at the close of the Triassic epoch, and now termed the Rhætic formation, the strata are crowded with fossil insects. From this time forth the geologist never loses sight of the mammalia, and many deposits of later date contain a considerable number of species. In its fossil state, the *Microlestes* has been found both in Germany and England. However, time fails me to say what I have heard of the strange creatures which lived elsewhere, during the epoch when I was born. It is more than probable that the numerous gigantic birds, whose foot-prints are found in the Connecticut Valley, had *reptilian* affinities—just as, during the Oolitic period, the reptiles had *ornithic*, or bird-like affinities.

In South Africa there existed a peculiar group of reptiles termed *Dicynodonts*, from the peculiar walrus-like characters of their tusks or teeth. They occur there in such abundance that the strata can be identified by their remains. The dry land everywhere was covered by a flora resembling in many of its general characters that of the Carboniferous epoch. This is the last we see of the familiar coal forms, for others were already in existence, destined soon to replace them and render them extinct. Thus much, therefore, for the dim recollections of a piece of Rock-salt!

CHAPTER VIII.

WHAT THE PIECE OF JET HAD TO SAY!

> "And how the nuns of Whitby told
> How of the countless snakes, each one
> Was turned into a coil of stone,
> When holy Hilda prayed.
> Themselves, within their sacred bounds,
> Their stony coils had often found.
>   *   *   *   *   *
> "On a rock by Lindisfarne,
> St. Cuthbert sits, and toils to frame
> The sea-born beads which bear his name."
>
> SCOTT'S *Marmion*.

NOW few of the beauties, whose delicate ears, heaving bosoms, and supple wrists I am made to adorn, are acquainted with the faintest outline of my history and experience! I will leave it to my hearers to say whether my story is not worth listening to.

The period when I was born, and in whose rocks I am most commonly found, is that known to geologists by the name of the Lias. In the lignite portion of its strata, among the "Alum Shales," I occur in my natural state as lumps and nodules. When purest, I am deemed most valuable, on account of my use in the manufacture of the well-known jet ornaments. I am purely of *vegetable* origin—as much so as coal itself—although I am usually considered a species of "black amber." Like the yellow variety which goes by that name, I am electric when briskly rubbed. As a fossil pitch or gum, I am related to the peculiar coniferous flora which grew so abundantly, although in comparatively few species, during the Liassic epoch. The chief features of these vegetable forms I shall presently endeavour to describe to the best of my recollection.

First, let me say a word as to the rock formation in which I am found. · Why it is called the "Lias" few wise men know, so that I may be excused, seeing this name was given to it so many centuries after my birth. It is usually regarded as a corruption of the word "layers," and I think this is very probable, as the general appearance of the strata is such as to cause such a name to be given to them, *par parenthèse.* Thin bands of dark limestone alternate with equally thin bands of dark shale, like so many sandwiches; this "ribbon-like" arrangement is very persistent, at least in England, and from it may have come the name of "Lias," or "Layers." The modern science of geology includes in its technical list many names which had a humble origin among quarrymen and miners. However that may be, I well remember'the alternate stages of quiet and disturbance which affected the sea near where I was born. Sometimes its waters would remain calm and clear for years, during which colonies of shellfish or corals would grow over its bottom, and their

K

accumulated remains form a bed of limestone. And then the waters were thick and turbid with mud, which gradually settled to the bottom, lying on the top of the shell bed, and now appearing as a layer of shale. In fact, the alternation I have spoken of is itself a proof of the physical conditions which affected the Liassic sea. The thickness of the various strata is nothing like so great as that of the older formations, although the fossil remains are far more numerous, both in species and individuals. In the "struggle for life," which had been perpetually going on since the first appearance of life in the Laurentian epoch, many new forms had been developed. The total thickness of the Lias is only about eleven or twelve hundred feet, and this is usually separated into three divisions, termed respectively the Upper, Middle, and Lower. The upper portion consists chiefly of clays, whilst the middle is composed of "marlstone," crowded with fossils. This part is remarkable for its containing iron-ore in such abundance as to be worked for that valuable metal in some localities. The Lower Lias is that most characterized by partings of shale and limestone, already mentioned, and is by far the thickest member of the group.

The dry land of this period was broken into a series of undulations, as it is at present, although the mountains were not so high as they are now. The uplands were thickly covered with woods and forests of Araucarian pines and thickets of fern;

WHAT THE PIECE OF JET HAD TO SAY! 131

whilst the lowlands were green with densely-packed cycads, plants now confined to tropical regions. About one hundred species of Lias plants are known to science, but not one has yet been met with which belonged to the class of which the oak, ash, or nettle are familiar examples. Indeed, this group was not introduced until the Cretaceous epoch, which followed the Liassic, after the lapse of enormous periods of time. The ferns were remarkable for having reticulated veins traversing their fronds. In the damper places, and by the riversides, there grew miniature forests of equisetites, nearly allied to existing species. This was almost the only "English" feature about the Liassic landscape. The trees grew in many places on the lowlands by the sea, and the dark mud was often charged with the bituminous or resin lumps, which, under the name of "jet," now compose my personal substance. Amid this somewhat monotonous vegetation there lived several species of miniature marsupials—the only warm-blooded creatures then in existence—which found

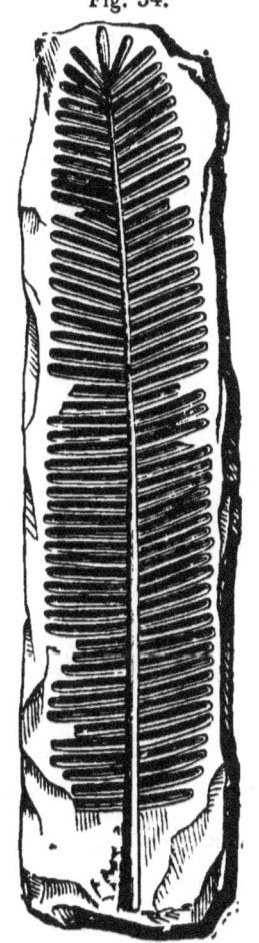

Fig. 54.

Fossil Fern (*Pterophyllum*).

the chief means of their subsistence in the hosts of insects that peopled grove and plain. Land reptiles, also, were not absent, both as crocodiles, tree-lizards, and flying-lizards.

This was, indeed, "the Age of Reptiles." Reptilian life was then modified to the various functions now fulfilled by a higher class—the Mammalia. In the air, on the land, in the water, one met with reptilian adaptations at every step. The places now filled by the whales and seals were then occupied by the *Ichthyosaurus* and the *Plesiosaurus*. The great land reptiles (*Deinosauria*), which became so abundant during the later—I may say the continuing "Oolite" period—stood in the room of modern carnivora and herbivora. Instead of bats and birds winging their way through the air, there were groups of *Pterodactyles*, some of them larger than the greatest bird now living. And, just as there is a *mechanical* and anatomical arrangement now characterizing the specialized mammalia, and thus fitting them for their various functions and places, so during the "Age of Reptiles" the relatively lower forms were built on the same plan. The modification which converts the limbs of a whale into flappers, also converted those of the *Ichthyosaurus* into paddles; the adaptation which provides a fulcrum for the muscles of a bird by fusing several bones together, we find applied to the flying-lizards of the Lias period. So wonderfully simple is the great plan on which the Creator has chosen

WHAT THE PIECE OF JET HAD TO SAY! 133

always to govern the development of organic beings.

Sometimes the lumps of resin which had oozed out of the pine-trees floated seawards, and were afterwards buried in the muds along the bottom. At others, the marsh lands where the woods grew were encroached on by the sea, and from terrestrial passed to marine conditions. It was whilst I lay thus that I formed my vivid impressions of the strange creatures which swam above me, and whose deceased bodies occasionally sank down into the mud to rest by my side, until I was rescued; in my mineral condition as "jet," by that complex being called "Man." I will endeavour to recall the most remarkable of these creatures. First there was the *Ichthyosaurus*, or rather, several species of that reptile. As its name implies ("fish-lizard"), it was modified to a purely marine life; which its deeply double convex vertebræ also indicate. Some of the larger individuals at-

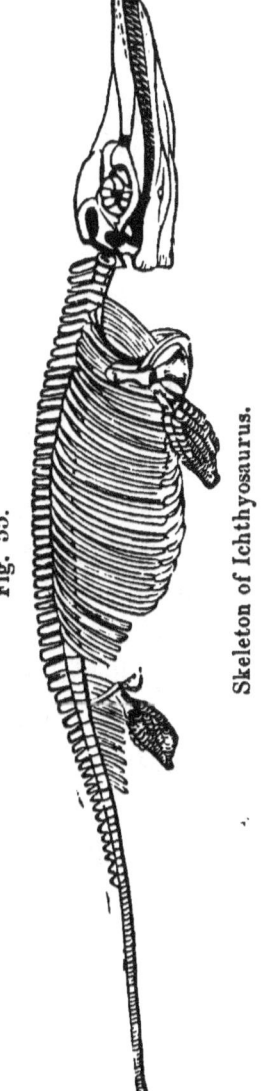

Fig. 55.

Skeleton of Ichthyosaurus.

tained a length of thirty feet, and I remember them going through all the usual routine of their reptilian life in the waters along whose floor I lay and watched. They were carnivorous in their habits, feeding on the larger fishes, and even on one another. To the best of my belief they differed from most reptiles in bringing forth their young alive. Many a time have I seen one of their carcasses, floating by means of the decomposed gases, right over where I lay; by-and-by the gases would escape, and the body sink to the muddy bottom; there it lay and was mineralized, and thence the geologist now disinters it in long ages subsequent to the elevation of this sea-bed into dry land. And his researches bear out the truth of what I say, for he frequently finds the fossilized remains of the reptile's last meal enclosed within the ribs where the stomach once lay, and even the fossil *fœtal* remains of its young within the pelvic cavities. The *Ichthyosaurus* was indeed the tyrant of the Liassic seas; its crocodile-like head was armed with scores of conical teeth, implanted in a continuous groove; the rest of its body was not unlike that of a small whale, having similar paddles and tail.

Still more nearly related to the Lizard family (as its name implies) was the *Plesiosaurus*, whose habits, however, were quite different from its more tyrannical congener. Its head was much smaller, although thoroughly reptilian, and terminated a long neck, not unlike that of the swan, but longer, for it sometimes contained as many as forty vertebræ; its teeth were implanted in sockets, like those of the

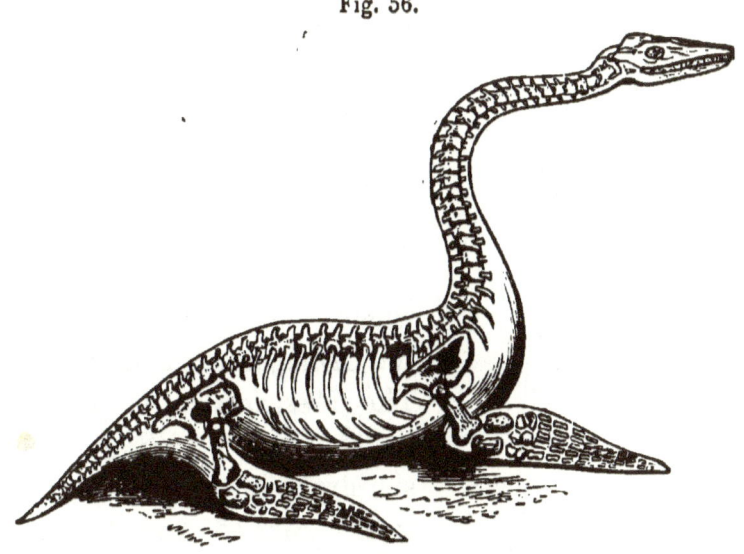

Fig. 56.

Skeleton of Plesiosaurus.

modern crocodile, so that, with a neck resembling a snake, a body and tail like those of a quadruped, and

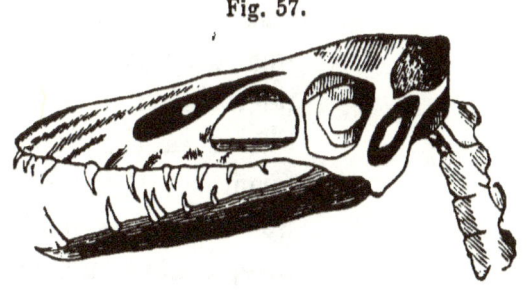

Fig. 57.

Fossil Head of Pterodactyle.

having paddles like the turtle, the *Plesiosaurus* had combined in itself structural adaptations now distributed among half a dozen widely separated animals. The largest of these queer-looking reptiles was twenty feet in length. Usually, its locality was by the seashore, in the shallower waters, where, by the aid of its long and flexible neck, it could dart at and seize the finny tribes as they swam past. It breathed air, as the whale does, and, indeed, as the *Ichthyosaurus* also did. The *Pterodactyle*, or winged lizard, was buried at sea simply because it was sometimes carried out by the wind, or else because its carcasses were carried seawards by the rivers; but it sometimes frequented the shallower mud flats on fishing expeditions. Anyhow, its remains were frequently buried in the deposits then forming. If the *Plesiosaurus* was a strange-looking creature, believe me, the *Pterodactyle* was much more singular. Some of the specimens must have been nearly fourteen feet across the

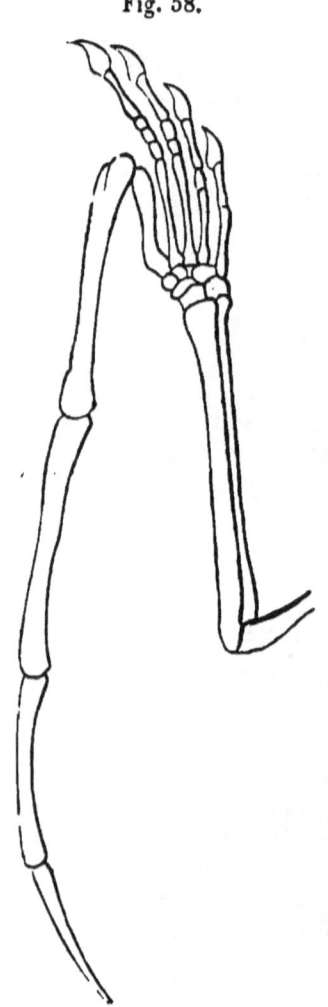

Fig. 58.

Fossil Skeleton of Wing of Pterodactyle.

spread of wings! Imagine a creature of this kind, possessing a long-snouted, crocodile-like head, and a long bird-like neck, with wings like those of the bat, a smallish body, and little or no tail! And yet this type of reptile did not depart from the

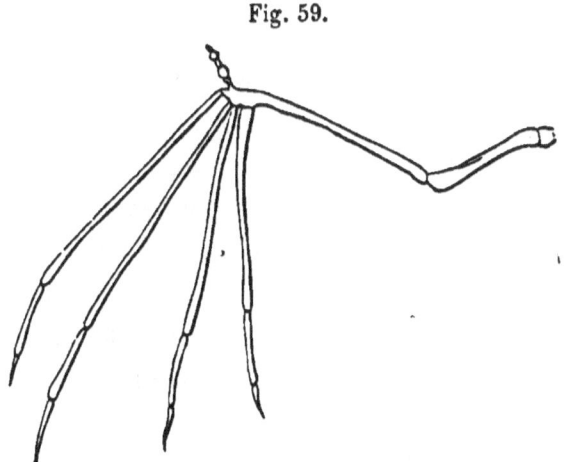

Fig. 59.

Skeleton of Bat's Wing.

normal form more than does the duck-billed Platypus from existing mammalia. The *Pterodactyle* could perch on trees, hang against perpendicular surfaces, stand firmly on the ground, hop like a bird, or creep like a bat.

So much for the reptiles with which necessity made me acquainted. I cannot speak much for the others, as most of them were not very common until later on. But the fish which lived in the Lias sea were almost as strange, compared with recent forms, as the reptiles. Most or all of them were covered with bony plates instead of scales, each plate being glossy with an enamelled varnish. Among the commonest of these fishes were the *Dapedius*, which had its scales set like a mosaic pavement—hence its name. The *Lepidotus*, or "bony pike," was related to a family still living in Africa and North America, and its haunt was usually off the mouths of rivers or in estuaries. The *Æchmodus* had a peculiar, "bream-like" appearance, whilst its small mouth was set with sharp, needle-like teeth. The *Acrodus* was a fish which lived on mollusca, &c., and its teeth were adapted for bruising and crushing them. In their fossil condition they go by the vulgar name of "fossil leeches," on account of the fine striæ which converge towards the centre of the upper surface. The *Hybodous* was a fish of altogether different structure, having shark-like teeth, and very formidable and well-developed spines on the dorsal fins. Hosts of smaller fry abounded, but my recollection does not go back so vividly towards them.

It would certainly be a gross mistake not to recall the appearance of one very remarkable object—the *Extracrinus*, or *Pentacrinus*, as it used to be called. This was the commonest of the Encrinites, which lived in the seas of the period. Of course my hearers are well aware that this object is nearly related to the "feather-star" (*Comatula*), which is anything but rare in British seas. But, instead of being free, as is the case with the latter object, the *Extracrinus* was usually fixed. Sometimes this was to drifting wood, but usually to the sea-bottom,

Fig. 60.

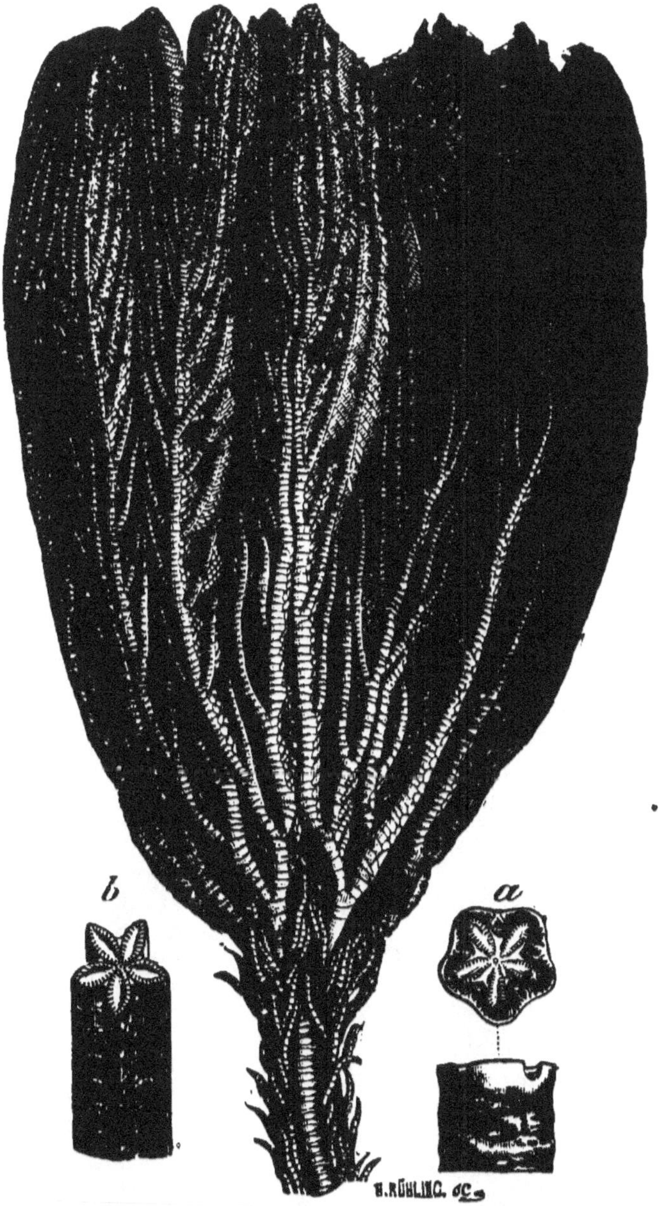

Pentacrinus fasciculosus. Lias Shales, Wurtemburg; *a* ossicle of stem (*b* ditto of another species, *P. basaltiformis*).

where it grew in thick submarine forests. In some places the Lower Lias shale is composed of hardly anything else than the remains of these fossils. Frequently they are changed into iron sulphite, or pyrites, and then they have a very brilliant appearance when first laid open with the chisel. This splendour, however, is very transitory, for the atmosphere plays sad havoc with them. The whole structure of the *Extracrinus* was built up of little ossicles, or joints, which fitted one into another, so that mobility as well as strength was obtained. The arms divided and subdivided into an infinite complexity, but all were arranged around the central mouth. One individual alone contained scores of thousands of joints or ossicles, like living nets. These complex arms groped through the waters in search of food. Nothing could be more graceful or elegant than the forms and motions of these extinct crinoids.

In many places the sea-bottom was a perfect aggregation of colonies of conchiferous shells. The *Ammonite* and *Nautilus* floated on the surface, and sometimes crept along the bottom. That strange-looking, cuttlefish-like creature, the *Belemnite*, swarmed in such numbers that the internal bones sometimes lay on the sea-bottom in hundreds. One species, at least, of the *true* cuttlefish lived along with them, for its ink-bag has been found fossilized and its ink so unexpended that the creature's likeness was drawn with it! The *Nautilus* was an old inhabitant of the world when the *Ammonite* was introduced on the stage of existence. As a family, it had reached the maximum of its existence, and was slowly waning into extinction, although it has been able to survive the flourishing class of *Ammonites,* for one species still

Fig. 61.

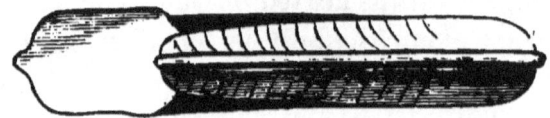

Ammonites bifrons.   Side view of ditto.

represents it! Seventeen European species of *Nautilus* are known from the Lias strata alone. But the *Ammonites* were by far the most abundant, and I, may say also, by far the most beautiful, of all objects

WHAT THE PIECE OF JET HAD TO SAY! 143

which lived at this time. Nothing could be more graceful and varied than the outward forms of different species. They differed in structure from the *Nautilus* in having the divisional chambers foliated along their edges, instead of being straight.* Another leading distinction was the position of the air-tube, or *siphuncle*, which did not run centrally through the chambers, as it did in the *Nautilus*, but along

Fig. 62.

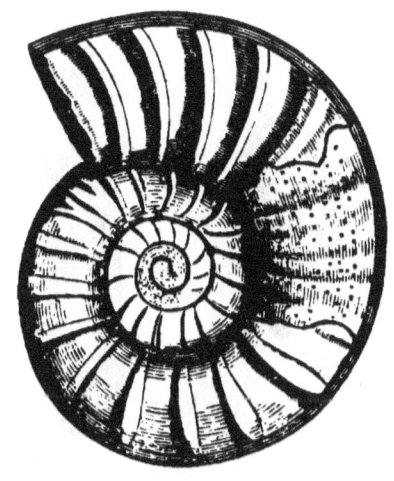 

Ammonites obtusus. Side view of ditto.

the margin of the outside, or back, of the shell. No fewer than 266 species of *Ammonites* are peculiar to the Lias deposits of Europe, whilst those of Britain

* The characteristic shell of this class which abounded in the deeper parts of the Triassic seas, where the limestones were forming, is the *Ceratite*. Its chambers are not so completely foliated as those of the *Ammonites*. Singularly enough, the young of the latter greatly resemble the *Ceratite* in this respect.

alone contain 128. Next in abundance to them were the *Belemnites*—vulgarly called " Thunder-bolts ".—above mentioned. The Lias strata of Great Britain have yielded 105 species, the British beds alone having produced 57 of them. The *Brachiopods*, or " Lamp-shells," which were so abundant during the Silurian and Carboniferous periods, were much more scantily developed in Liassic times. Here you see the last of the genus *Spirifer*. On the

Fig. 63.

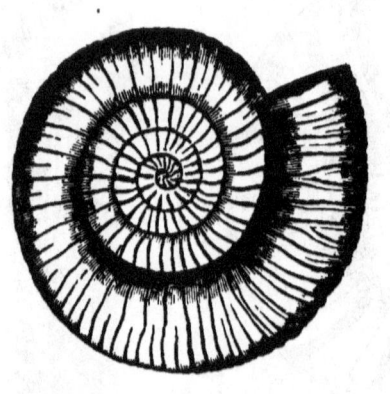

Ammonites communis.   Side view of ditto.

other hand, the true *conchiferous* species, which had lain in the background during the earlier epochs of our planets' history, now began to assert that supremacy which they still hold in even a greater degree. No fewer than 625 species of Conchifera have been found in the European Liassic deposits alone. The commonest among these were the species of *Gryphæa*—a kind of curved fossil oyster, whose abundance sometimes makes up entire beds of limestone. The

WHAT THE PIECE OF JET HAD TO SAY! 145

Hippodium, Plagiostoma, and *Avicula* are also very common. Of brachiopodus shells, including such familiar types as *Rhynchonella, Terebratula,* &c., there are as many as 115 species peculiar to the Lias strata of Europe. Taking the summary of fossils which have been found in the strata of this age in Britain, including plants, insects, shells, and vertebrata generally, there are no fewer than 1228 species known to science. This, of course, is not all; for the list of known species has been more than doubled within the last twenty years. It belongs to the science of the future to develop the fauna and flora of each period of the past, but I am firmly convinced that its efforts will be only to prove the continuity of the great Life-scheme, whose broken fragments are enclosed in the rocks. And yet, broken and shattered though they be, they are capable of being so put together that man—the last and highest link of the series—is able to spell out the grand plan of Creation, and to turn with mingled feelings of awe and admiration towards its Great Designer!

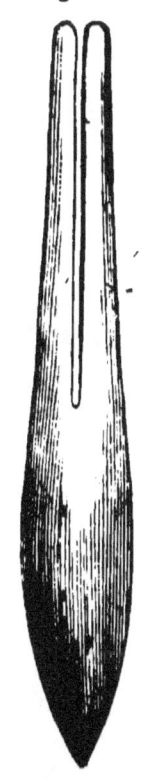

Fig. 64.

Liassic "Thunderbolt" (*Belemnites hastata*).

CHAPTER IX.

WHAT A PIECE OF PURBECK MARBLE HAD TO SAY!

"Contemplate all this work of Time,
   The giant labouring in his youth;
   Nor dream of human love and truth,
As dying Nature's earth and lime;

"But trust that those we call the dead
   Are breathers of an ampler day
   For ever nobler ends. They say
The solid earth whereon we tread

"In tracts of fluent heat began,
   And grew to seeming-random forms,
   The seeming prey of cyclic storms,
Till at the last arose the man."
                            In Memoriam.

       *   *   *   *   *

"On the pavement lay
Carved stones of the abbey-ruin in the park,
Huge Ammonites, and the first bones of Time."
                                    The Princess.

THERE are few of my intelligent hearers who are not acquainted with the peculiarities of my appearance. In this civilized country, where old churches abound, I may have formed a portion of the fonts in which they were christened, or the pillars of the Early English doorway by which they will be carried to receive the last sacerdotal rites. As a slab near the

Fig. 65.—Ideal Landscape of the Oolitic Period.

altar, some of them may have stood on me whilst they took upon themselves the solemn duties of matrimony, little dreaming of the long lines of generations the obscure stone at their feet could tell them about!

I belong to the upper part of that geological formation termed the "Oolitic," from the peculiar "roe-like" appearance often presented by some of its limestones. This general name is another of those instances of the early nomenclature of geology which are obliged to be retained now from their extended use, although they are found to be no longer specially applicable. Of course I cannot be expected to remember exactly what took place before I was born; all I can do is to tell you what I have heard, handed down by oral tradition through the long line of my ancestors. I am the last of the family, and left no descendants. After me came that series of deposits included under the general term "Cretaceous" or Chalk. But, as my hearers would expect, there are palæontological reasons for myself and brethren being grouped together. These are chiefly the family likeness of our included fossils, marine, fresh-water, and terrestrial. I heard what my cousin the Piece of Jet had to say, and may here remark, that it is a pity his formation is not considered as one of us, and not treated as if he were simply a distant connection. Many of his fossils are so much like those of our family that, even if they are specifically distinct, a good relationship to us may be made out of them.

The lowest beds of the great geological system to which I belong go by the modern name of the "Inferior Oolite." But though these follow in direct order, there was a great interval of time between the succession. This is plainly shown by the fact that out of the hundreds of species of fossil shells peculiar to the upper parts of the Lias, not quite forty species lived long enough to become fossilized in the lower beds of the Oolite; many of the rest became extinct, while others perhaps migrated to areas where the physical conditions better suited them. There was a greater longevity in certain creatures then, just as there is now; for we find several species of bivalves and ammonites existing during the long period of time which elapsed whilst the entire series of beds composing the Oolitic formation was being slowly deposited.

I will just give you the list of the principal of this series, mentioning them first in the order of their antiquity or seniority—a practice no doubt in vogue among yourselves. After the Inferior Oolite comes the Great, or Bath Oolite, and Stonesfield Slate. The Cornbrash and Forest Marbles complete what is termed the "Lower Oolite." Then come the Oxford Clay and Kelloway Rock, both perhaps contemporaneous—the Coral Rag completing the "Middle Oolite." The Kimmeridge Clay, Portland Stone, and Purbeck series form the "Upper Oolite," and bring the entire formation to a conclusion. These deposits stretch across England, in a belt, of about thirty miles in width, from Yorkshire to Dorsetshire. They follow each other in tolerably regular order, and as they are relatively composed of shales, sandstones, and hard limestones, and as the entire series has been much exposed to atmospherical and marine wear and tear since they were solidified and upheaved, it follows that this denudation has been so operative as to wear away the softer beds and to leave the harder standing. Hence the physical geography of the whole formation differs according to the underlying geology. Deep valleys

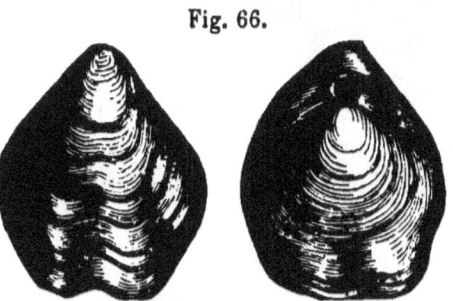

Fig. 66.

Terebratula biplicata—a common fossil.

or extensive plains lie where the clayey or argillaceous strata crop out; and broken hills, frequently with more or less steep westerly escarpments, indicate the areas occupied by the limestones and harder sandstones.

As might be expected, when it is remembered that this series of deposits was formed chiefly along the old sea-bottoms, there must have been an extensive and long-continued list of geographical changes rung whilst it went on. The bed of the ocean was alternately the receptacle for the fine muds brought down by rivers, along whose deltas grew the rich vegetation locked up in the coal-seams and shales of the Lower Oolite near Scarborough. Then we have evidence of a depression of the area, which removed the sphere of deposition of the mud, and brought clear water over the site. Here the physical conditions allowed mollusca, corals, &c., to swarm in abundance, and their accumulated remains

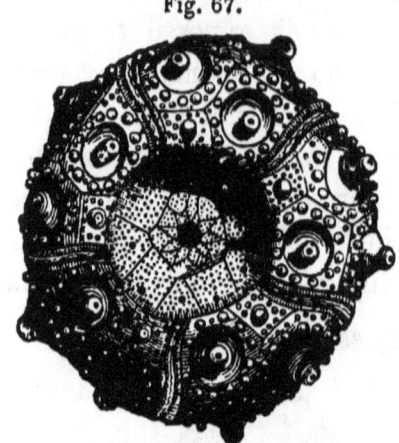

Fig. 67.

Cidaris coronata—a common Oolitic echinoderm.

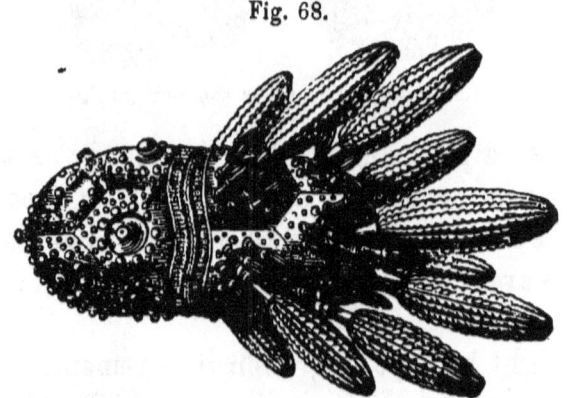

Fig. 68.

Ditto, showing the spines attached to tubercles on right hand.

thus formed the limestones. Calcareous sandstones were formed out of the comminuted coral reefs, shells,

WHAT A PIECE OF PURBECK MARBLE HAD TO SAY! 153

&c. Occasionally, influxes of mud killed off large numbers of encrinites, as at Bradford, near Bath, and buried them beneath its *débâcle*, clear water returning shortly afterwards, as the parasitic zoophytes, &c., which attached themselves to the broken joints of the encrinites, plainly indicate. At length the deposits more or less filled up the shallower parts of the sea, and upheaval converted a portion of it into dry land. The hollows of this land became fresh-water lakes, in which swarms of *Planorbis*, *Paludina*, and other well-known fresh-water snails lived. The water was clear, and there was no great amount of muddy materials carried into these lakes. Time only was required for the shells to accumulate along their floors to such an extent, that, in their solidified condition, they form the bulk of that well-known "Purbeck Marble" of which I am a humble and minute portion. Occasionally the sea-waters backed up the fresh, and encroached on some portion of the lakes, holding the place sufficiently long for brackish-water shells to live and multiply there, and to leave their remains behind them in token of what I have said. Even the pure sea-water once or twice gained ground, as the beds of fossil oysters, &c., intercalated in the Purbeck beds reasonably show us. In these different beds you find evidences of nearly all kinds of deposition, from the tolerably

Fig. 69.

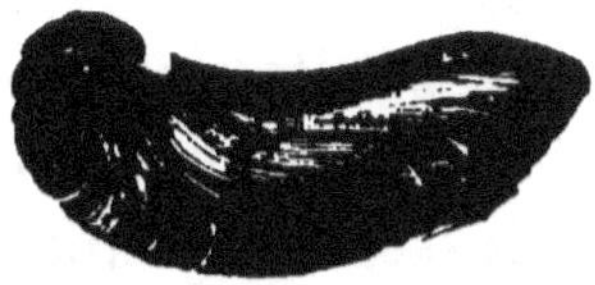

Gryphea—a common Oolitic and Liassic fossil.

deep water in which the "Coral Rag" was formed, chiefly as a coral reef, to the ripple-marked flagstones of the "Great Oolite," in which also you get tracks of worms, crustaceans, &c. The total thickness of the entire series is about two thousand four hundred feet, which alone will give you some idea of the enormous period of time represented by them.

There are few geological formations so rich in fossils as the Oolite. Not only in individuals, but

Fig. 70.

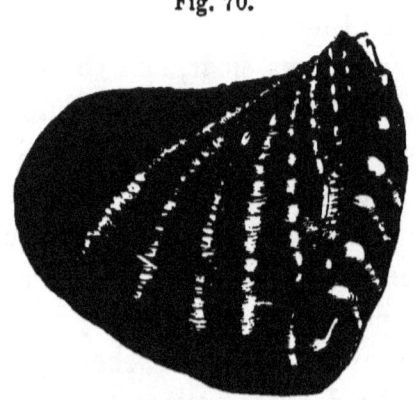

Trigonia—a common Oolitic fossil.

also in species, the rocks are one vast museum, illustrating a particular stage in the world's past history. You may catch glimpses of life in every form of its enjoyment—in the mighty Saurians which frequented the open seas; in the busy coral reefs secreting lime; in the bony-plated fishes, whose enamelled scales glanced through the waters. You see the low tide fringed by a vegetation, partly growing on the mud-banks as a swamp, and you distinguish forms now regarded as sub-tropical to Britain. The sea-bed is literally alive with cidarids, bivalves, univalves, sea-lilies, and lamp-shells. Overhead, over land and water, the flying-lizards (*Ptero-*

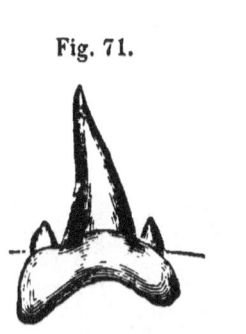

Fig. 71.

Fossil Oolitic fish tooth, nat. size.

Fig. 72.

Section of tooth magnified.

dactyles) whirl and swoop. The tiny kangaroo rats and opossums are busy in the forests, some lying in wait for their numerous insect prey, and others, more bloody-minded, are cannibally inclined! The great fresh-water lakes, along whose floors I was formed by the accumulation of ordinary fresh-water shells, were set in a dense and beautiful framework of pine-trees, of cycads, zamias, and tree-ferns. But, vast as the period of time is since this, the last of the Oolitic series, was formed, numbering, as it undoubtedly does, *millions* of years, it has all elapsed within the lifetime of existing genera of shells! The *Paludina*, which principally make up my bulk, can hardly be told, even by experienced conchologists, from the ordinary fresh-water snails which still inhabit English rivers! In structure of limb, teeth, and general adaptation, the highest orders of animals then existing were wonderfully like their Australian brethren.

In the swampier places, at the beginning of the Oolitic period, where the vegetation grew thick and rank, beds of peat were formed and covered up by mud. This peat subsequently became coal. The iron

Fig 73.

Phlebopteris—a characteristic Oolitic fern.

diffused through the muddy mass was influenced by chemical action, so as to reunite and segregate, as an argillaceous carbonate, into layers and nodules of *iron-stone*. In this respect, the physical condi-

WHAT A PIECE OF PURBECK MARBLE HAD TO SAY! 157

tions greatly resembled those which existed during the Carboniferous epoch, and therefore the results are very similar. All you have to do is to transpose the animals and plants of the two eras, the difference in each of which represents the amount of time which had elapsed between them, and in which the vital modifications had taken place. In the

Fig. 74.

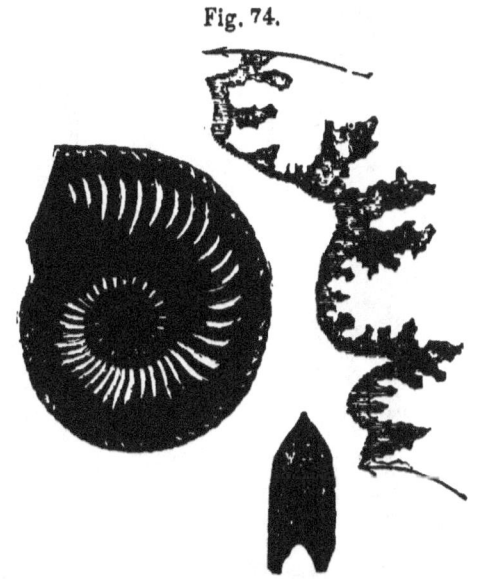

Ammonites amaltheus, with sketch of foliation of chambers.

Stonesfield slate—a calcareous shale, and a capital burial-ground of extinct animals—there were entombed the remains of at least four species of mammalia. As I before remarked, however, all the warm-blooded animals which lived during the Oolitic period belonged to the lowest order of their kind— the marsupials, or pouched animals, notorious for

158 WHAT A PIECE OF PURBECK MARBLE HAD TO SAY!

bringing forth their young in a half-gestated condition. When you ascend higher in the Oolitic series, through the more purely marine deposits (where,

Fig. 75. Fig. 76.

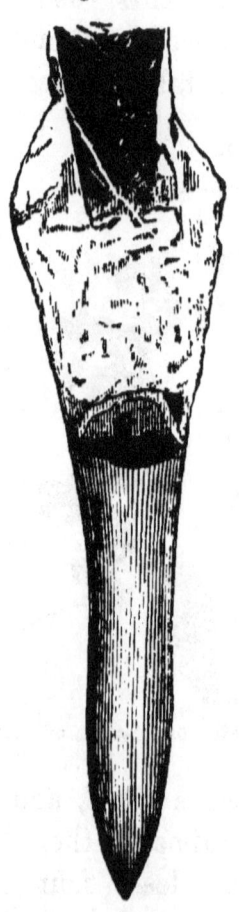
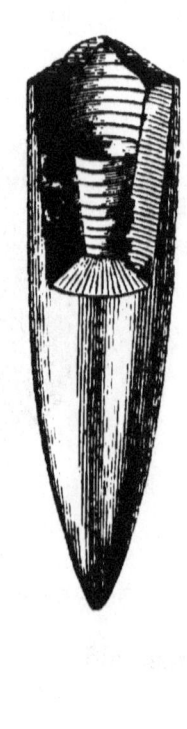

Oolitic " Thunder-bolt " (*Belemnites puzosianus*).   Oolitic " Thunder-bolt " (*B. abbreviatus*).

of course, you would not expect to find the land creatures well represented), and come to the Purbeck

WHAT A PIECE OF PURBECK MARBLE HAD TO SAY! 159

beds, then you will be astonished at the large number of species of marsupials, and the great modification and adaptation in their habits which had taken place. The streams entering the lakes where the Purbeck marble was formed were much more

Fig. 77.

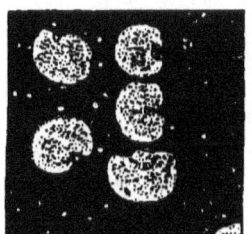

Oolitic Wood × 4½.

Fig. 78.

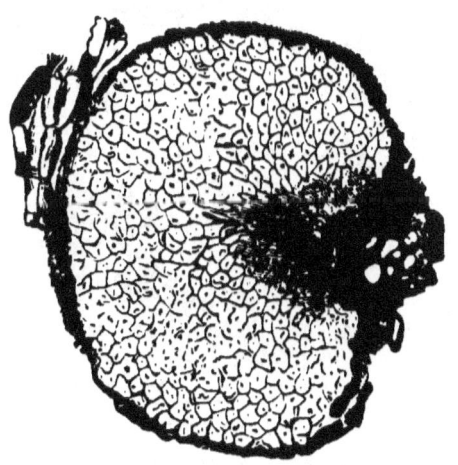

Oolitic Wood × 30.

likely to carry the carcasses of these dead marsupials there, and therefore the bottom of that lake was more likely to be a richer cemetery of their remains.

160 WHAT A PIECE OF PURBECK MARBLE HAD TO SAY!

Some of the Oolitic strata are much more favourable to the preservation of organic remains than others, and these invariably give us a glimpse of animal and vegetable life which, although of a

Fig. 79.

Microscopic sections of Fossil Wood, from Scarborough × 100.

much lower organization on the whole than the present, was admirably adjusted each to the other. Thus, the fourteen species of marsupials above mentioned, were all obtained from a thin seam, three or four inches thick, in the Purbeck series, and from an excavated area of about five hundred square yards! Of all these rich fossiliferous deposits, however, perhaps the most interesting is at Solenhofen, where there occurs the stone of that name, much in use now, I am told, for lithographic purposes. The sediment of which it is composed is very fine, so that the quality which gives it its economical value to man is exactly that which has rendered it such a splendid mausoleum for the fossils of the Oolite. Forty years ago there had been obtained from this one deposit no fewer than between two and three hundred species of fossils, of which seven species were those of flying-lizards, or *Pterodactyles;* six species were those of huge saurians; three were tortoises; sixty species were fish; forty-six were crustaceans; and twenty-six were insects, which had probably been blown from the land by the breezes, and eventually found a watery grave, and an immortality they never dreamt of.

I have already spoken a little of the peculiar vegetation of this period—of the Cycads and Zamias and Tree Ferns, which had taken the place of the Calamites, Sigillarias, and Lepidodendra of the Palæozoic epoch. Besides these, there flourished other plants, now regarded as characteristically Australian, of which the Araucarian pines are examples; several species are found in the Inferior Oolite, whose cones showed that they lived and flourished not far distant. Then, again, in the so-called "dirt-beds" of the

Portland stone, and also of the Purbeck beds, you have evidences, not only of old land surfaces, but also of the dense vegetation which covered them. These "dirt-beds" plainly indicate the extended period during which these old cycadean and pine forests grew. Their remains are now found silicified, their trunks and stems lying recumbent amid the "dirt," whose fresh-water shells tell you how it had been the shallow bottom of a lake before it was a forest-bed, and that it was there its rich black soil was accumulated! The Cycads are flattened somewhat by the pressure of the overlying beds, so that their bracts or scales give them a peculiar appearance, which, I am told, has earned for them among the quarrymen the name of "Crow's Nests."

Fig. 80.
Cycadoideu, or "Crow's Nest."

As you are perhaps aware, the sea was still the home of the great fish-lizards, *Ichthyosaurus, Plesiosaurus*, &c. On the dry land the reptile family was represented by an abundant group, which goes under the general name of *Deinosauria*, or "terrible reptiles." Judging by the size of some of them, this name is not badly given. But by far the most characteristic feature about these huge land reptiles was their near anatomical relationship to the *birds!* You hear a good deal of foolish talk now about

WHAT A PIECE OF PURBECK MARBLE HAD TO SAY! 163

"missing links," and those who make use of it little know that all the fossils are, more or less, of this nature, and fill up gaps in the natural history classification. Some of the reptiles of which I am speaking walked on two legs, like great Cochin China fowl, and with their hind quarters much more strongly developed than their fore limbs. In this respect they resembled, amongst the reptilia, the position of the kangaroo, which, as everybody knows,

Fig. 81.

Megalosaurus—a great terrestrial reptile.

generally uses only his huge hind legs, his fore limbs being much smaller and weaker. One of these land reptiles, named *Compsognathus*, whose remains have been found in the Stonesfield slate, and which was only about two or three feet in length, is the nearest approach, in its general structure, to birds of any yet made known. As you are aware, nearly all reptiles are egg-bearing in their habits, and the fossil eggs of the Oolitic reptiles have been met with, showing that, so long ago as the Oolitic age, this class had the same habits as their diminutive representatives of the present day. But what is very remarkable is, that whilst the reptiles of this period had *bird-like* characters, the only birds known had *reptilian* pecu-

Fig. 82.

Archæopteryx (restored), from the Oolitic Limestone of Solenhofen.

liarities! No doubt you are aware that these two great groups of animals, birds and reptiles, follow each other in ordinary classification. They do so *in order of time*, the reptiles first in their lowest grade as Amphibia (*Labyrinthodonts*), which gradually rise to a higher standard in the Ichthyosaurus, until they assume features which, as I above remarked, now belong wholly to birds. Singularly enough, the true birds follow soon after, and the first specimen you meet with shows, in the structure of its tail-bones, &c., that it had borrowed some of the anatomical peculiarities of the reptiles! This bird had a long attenuated tail, like that of a lizard, with feathers bifurcating from each side down to the end. This strange bird is now known as the *Archæopteryx*, and its bones, and even *feathers*, have been found beautifully preserved in the Solenhofen stone. Here you have, at any rate, a meeting-ground on which two of the great divisions of the animal kingdom exhibit their mutual descent. It is a suggestive fact for those of my hearers who are sceptical about "missing links!"

The ages which have passed away since these things occurred are bewildering to those who are anxious to know, in so many years, how old the world is, as if that fact would add anything material to their real knowledge. At the time of which I am speaking, the area occupied by the Himalaya Mountains was a deep sea-bottom : that great mass has been slowly elevated to its present great height since the era of my birth. The Jura Alps were in the same condition, and have undergone similar elevation. One generation of animals and plants after another has passed away from the earth, having been slowly pushed out of existence by newly-formed species, better fitted to the alterations effected through the changes in physical geography. The whole of the Oolitic strata of soft sands, oozy lime, and dark mud, as well as the beds of loose freshwater shells, have undergone chemical action and change, and been transformed into sandstones, limestones, shales, and Purbeck marbles. Our family has been in past times, and is now, a favourite with man in his endeavour to express his religious convictions and æsthetical feelings. We form the stonework of his grand churches and cathedrals, and I myself have the honoured position of forming part of his altar, his christening-font, or his grave-slab! The tread of many generations of men has not effaced my lacustrine origin. Dynasties and religions have passed away, and been replaced by others breathing a more Christian and liberal spirit, just as the Oolitic animals were replaced by those of a higher organization; but I still form part of these grand structures, silently testifying to the endurability of nature over art, and yet myself a testimony that Nature herself is full of changes, and restlessly advances to a more perfect condition!

CHAPTER X.

THE STORY OF A PIECE OF CHALK.

"There rolls the deep where grew the tree,
Oh earth, what changes hast thou seen!
There, where the long street roars, hath been
The stillness of the central sea."

TENNYSON.

IT is so long ago that I can hardly remember it. My first recollections are of a white, muddy sediment, hundreds of feet in thickness, stretching along the bottom of a very deep

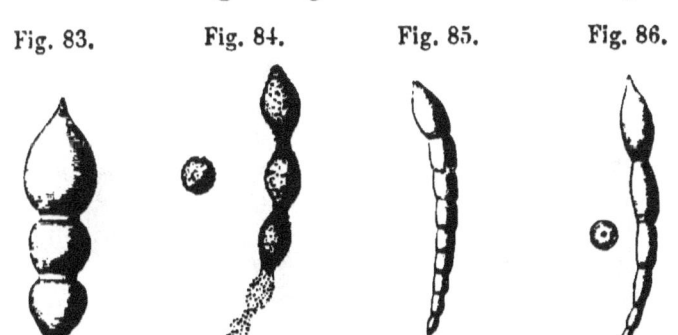

Fig. 83.   Fig. 84.   Fig. 85.   Fig. 86.

Nodosaria limbata, Chalk, Charing.   *Dentalina aculeata*, Chalk, Ilminster.   *Dentalina gracilis*, Gault, Folkstone.   *Dentalina Lorneiana*, Chalk, Kent.

FORAMINIFERA FROM THE ENGLISH CHALK.

sea. Of this oozy bed I formed an inconsiderable part. The depth of sea-water which pressed down this stratûm was so great that the light scarcely found its way through the green volume. Day and night the billows tossed and heaved above me. I

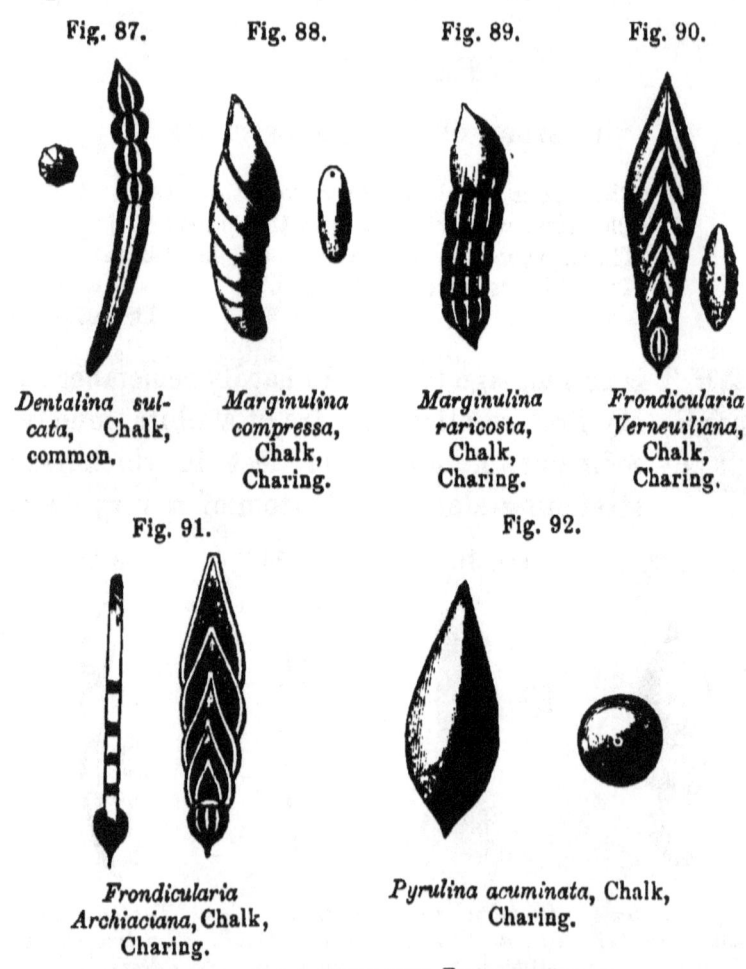

Fig. 87. *Dentalina sulcata*, Chalk, common.
Fig. 88. *Marginulina compressa*, Chalk, Charing.
Fig. 89. *Marginulina raricosta*, Chalk, Charing.
Fig. 90. *Frondicularia Verneuiliana*, Chalk, Charing.
Fig. 91. *Frondicularia Archiaciana*, Chalk, Charing.
Fig. 92. *Pyrulina acuminata*, Chalk, Charing.

FORAMINIFERA FROM THE ENGLISH CHALK.

could hear the storm howl and the hurricane sweep over the surface of the sea, although they could not affect the bottom where I was lying. Before I awoke to consciousness in my oozy condition, I had existed in quite another form. The constant beatings of the Cretaceous sea against its barriers, and, more particularly, the vast quantity of mineral matter poured into it by tributary rivers, had caused

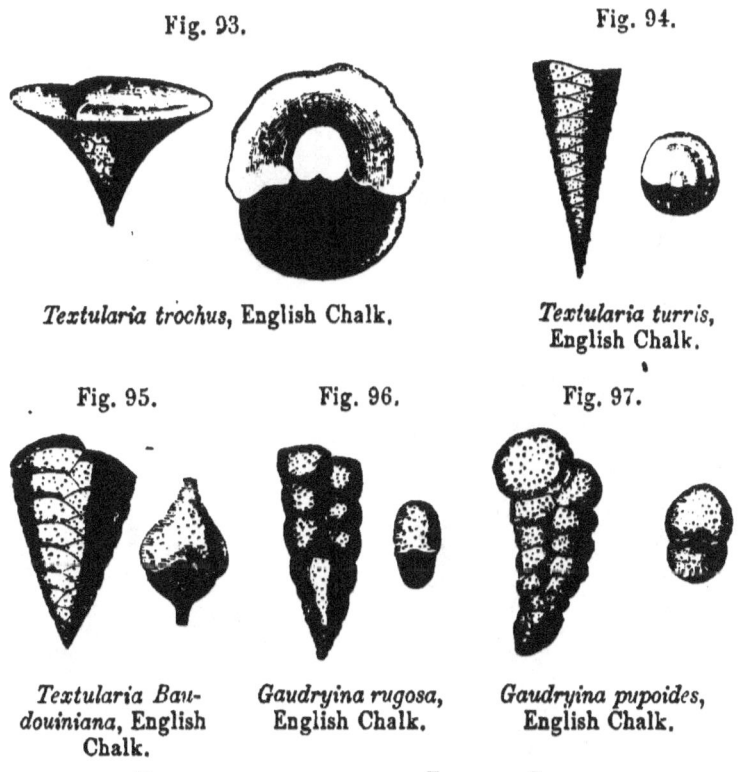

Fig. 93.   Fig. 94.

Textularia trochus, English Chalk.   *Textularia turris*, English Chalk.

Fig. 95.   Fig. 96.   Fig. 97.

Textularia Baudouiniana, English Chalk.   *Gaudryina rugosa*, English Chalk.   *Gaudryina pupoides*, English Chalk.

FORAMINIFERA FROM THE ENGLISH CHALK.

to be distributed through the sea-water a considerable quantity of mineral matter. Of course, great though this quantity originally was, when diffused through the sea, it appeared so small as not to affect the real transparency of the water. The presence of carbonate of lime (for such was a good portion of the mineral matter above mentioned) could only have been proved by delicate chemical tests. It happened, however, that there were organs sharp

Fig. 98.   Fig 99.

Verneuilina tricarinata, Kentish Chalk.   *Bulimina obtusa* (rare).

Fig. 100.   Fig. 101.

 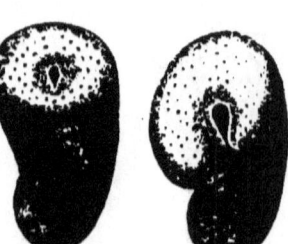

Bulimina obliqua, English Chalk, common.   *Bulimina variabilis* (rare).

FORAMINIFERA FROM THE ENGLISH CHALK.

enough to detect even this small modicum. These belonged to a group of animals so minute that you could have put millions of them into a schoolgirl's thimble!

Each creature was a perfect animal nevertheless.

It had a soft, jelly-like substance, which developed itself into feelers, that took hold of prey even smaller than itself. This soft body was enclosed in a sort of shelly case, beautifully ornamented, and

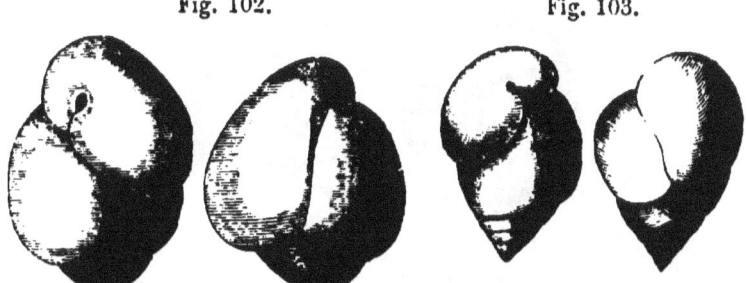

Fig. 102. Fig. 103.

Bulimina brevis, English Chalk.   *Bulimina Murchisoniana* (rare).

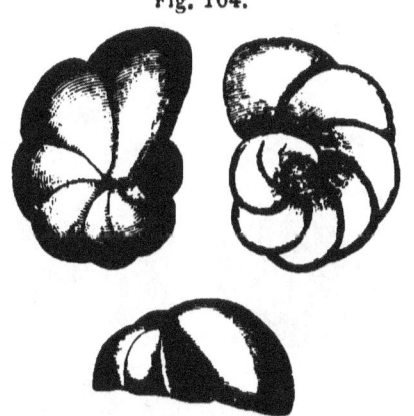

Fig. 104.

Truncatulina Beaumontiana, Gravesend and Warminster.
FORAMINIFERA FROM THE ENGLISH CHALK.

uniformly shaped. This case was manufactured out of the carbonate of lime, which has already been mentioned as held in solution by the sea-water. Every cubic inch of water in all the vast ocean at whose bottom I was lying, was alive with these animalcules, everlastingly at work separating the mineral matter. It was quite impossible to see these little workers

Fig. 105.

Rosalina Lorneiana (rare).

Fig. 106.

Rosalina Clementiana, Kentish Chalk.
FORAMINIFERA FROM THE ENGLISH CHALK.

that "out of water brought forth solid rock," and yet they were there. Their individual lifetime was very brief, rarely extending over a few days. But their

THE STORY OF A PIECE OF CHALK. 173

powers of reproduction were enormous, and thus they were always dying and generating. As they died, they began to sink slowly through the water.

Fig. 107.

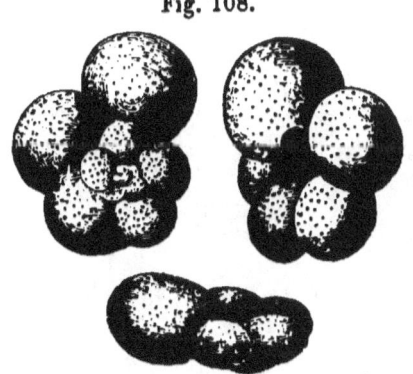

Rotalina Cordieriana, English Chalk.

Fig. 108.

Globigerina cretacea, English Chalk, common.
FORAMINIFERA FROM THE ENGLISH CHALK.

The sea was always full of their dead shells, which were gravitating towards the bottom, where they fell as lightly as the motes which float in the sun-

174   THE STORY OF A PIECE OF CHALK.

beams drop upon the floor. Night and day, they were always alighting there, and forming a thin film. Century after century passed away, and still found these dead shells accumulating, until all the figures I have heard reckoned on the black-board near me—I am now used in a school-room for the

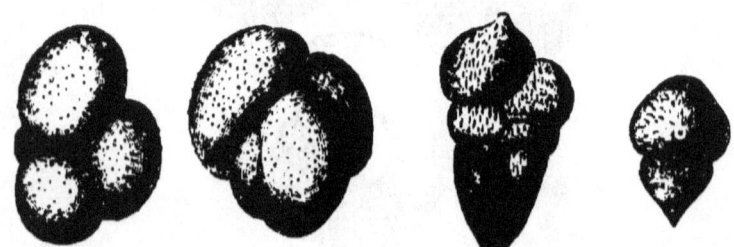

Fig. 109.

Globigerina elevata (rare).   *Sagrina rugosa*, Chalk, Charing.

Fig. 110.

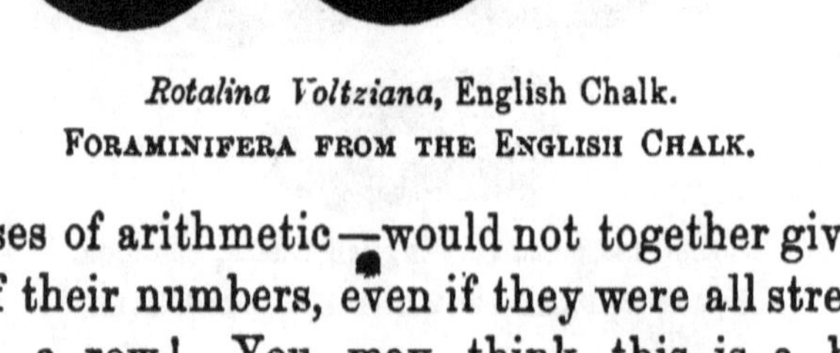

Rotalina Voltziana, English Chalk.

FORAMINIFERA FROM THE ENGLISH CHALK.

purposes of arithmetic—would not together give any idea of their numbers, even if they were all stretched out in a row! You may think this is a bit of romancing, but it is not. A few days ago, a gentleman broke a piece off me, and after powdering it and washing it with a fine camel-hair brush in distilled water (so as to make sure of his experiment), I heard him tell a friend that he could show him

Fig. 111.

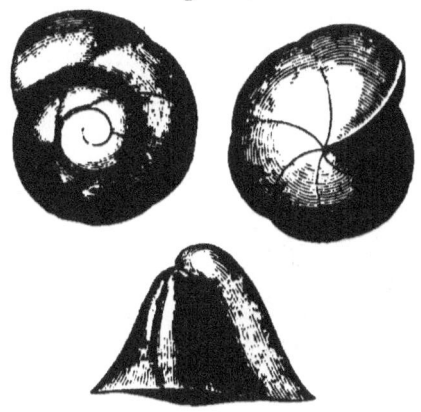

Rotalina Micheliniana, English Chalk.

Fig. 112.

Rotalina umbilicata, Chalk, Gravesend; Gault, Folkstone.
FORAMINIFERA FROM THE ENGLISH CHALK.

thousands upon thousands of fossil animalcular shells which he had obtained from this small piece! These minute shells belong to many genera or kinds, of which one called *Globigerina* is by far the most abundant. This genus lived along the bottom of the sea, and did not move, but gradually and continually secreted carbonate of lime from the water, so that in this way

Fig. 113.   Fig. 114.

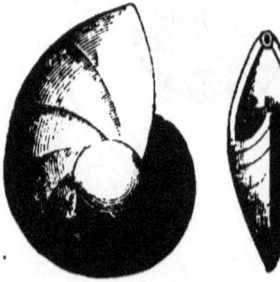 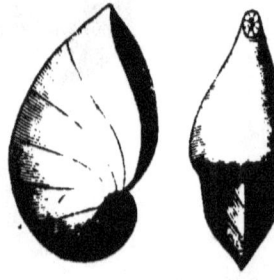

Cristellaria rotulata, variety.   *Cristellaria navicula*, Kentish Chalk.

Fig. 115.   Fig. 116.

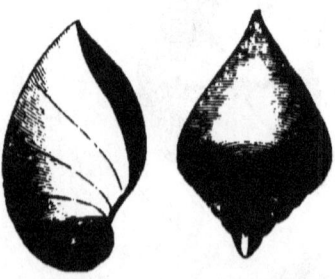 

Cristellaria triangularis, Chalk, Kent ; Gault, Folkstone.   *Cristellaria recta*, Chalk, Charing.

FORAMINIFERA FROM THE ENGLISH CHALK.

alone the limey mud would have increased in bulk, just in the same way that coral reefs grow at the present time. Although I am no bigger than a small orange, I can assure you there are scores of millions of fossil shells contained within my bulk. In fact, I am myself little more than a mass or congeries of the dead shells to which I before alluded. Every time the teacher makes a figure with me on the black-board, he leaves thereon thousands of fossil

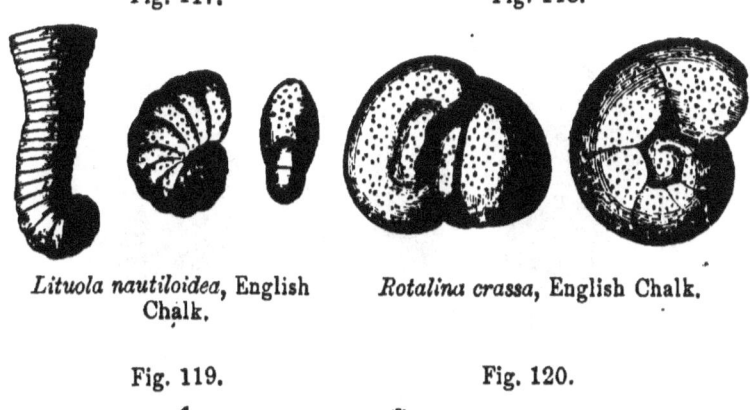

Fig. 117.

Fig. 118.

Lituola nautiloidea, English Chalk.

Rotalina crassa, English Chalk.

Fig. 119.

Fig. 120.

Frondicularia tricarinata, Kentish Chalk.

Flabellina rugosa, Kentish Chalk.

FORAMINIFERA FROM THE ENGLISH CHALK.

animalculæ. If you will wash the chalk as the above-mentioned gentleman did, you may see these minute fossils for yourself; though, it is true, you would need a powerful microscope to enable you to do so.

It was the gradual accumulation of these ani-

178   THE STORY OF A PIECE OF CHALK.

malculic shells that formed the oozy mud at the bottom of the sea. The extent of this mud-bed was very great—not less than scores of thousands of square miles in area. Notwithstanding the slowness of the deposition, and the infinitely minute creatures

Fig. 121.

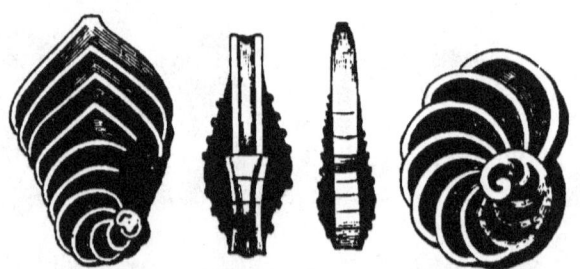

Flabellina Baudouiniana, English Chalk.

Fig. 122.   Fig. 123.

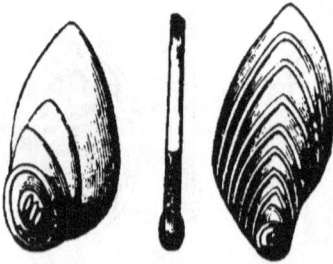 

Flabellina pulchra, Kentish and Norwich Chalk.

Cristellaria rotulata, English Chalk and Green-sand, common.

FORAMINIFERA FROM THE ENGLISH CHALK.

which almost wholly formed it, the accumulation went on until the mud had reached a vertical thickness of fifteen hundred feet! What must be the enormous number of shells contained in this mass, and the number of centuries occupied in elaborating it, I leave you to guess. The rate of deposition was very regular, and I have heard that along the bottom of the great ocean called the Atlantic, there is actually now being formed a stratum very similar to that from which I was taken. Like it, also, it is formed principally by immense numbers of dead animalculæ. The same species of *Globigerinæ* is still living along the Atlantic sea-floor, and I am told, is still engaged in forming similar chalky ooze to that which its lineal ancestors formed countless ages ago! Minute grains of chalk, of vital origin, are also abundant in my substance, and these, I am further informed, under the name of *Coccoliths*, are in equal abundance in the Atlantic mud, where they exist under the lowest forms and types of animal life.

I lay along the bottom of the Cretaceous sea for thousands of years, during which great changes took place in the oozy deposit, some of which I distinctly remember. I should have said that, besides carbonate of lime, there were diffused through the seawater other minerals, among the rest, one called *Silica*, the basis of common sand. Well, some proportion of the minute animals inhabiting my native sea may have used this mineral instead of lime, so that their shells were formed of flint. These, of course, fell to the bottom along with the others, and were all mixed up together. By-and-by, a chemical change took place in the thick mud. It seems that the little grains or shells of silica have a tendency to separate from the lime, and to run together;

consequently, the flinty little shells aggregated along the sea-bottom, and there formed what are now known as *flint-bands* and *nodules*. A chemical process, resulting from the decomposition of animal matter, caused some of the dissolved silica to be precipitated, and thus hastened the formation of flints. These layers of flint were formed at nearly regular intervals, the chemical changes being very

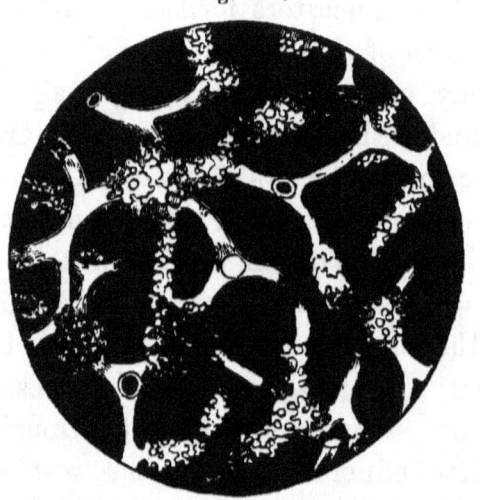

Microscopic Section of *Polypothecia*, a fossil siliceous sponge, magnified 150 diameters. From Green-sand, Carrow Well, Norwich.

uniform. Again, another, and perhaps a principal, means of forming *flint* was by the decomposition of the animal matter, which was the means of precipitating the silica held in solution by the sea-water. I should also mention, that as the oozy bed increased in thickness, what with the weight of sea-water and the overlying mud, the *lower* beds began to be com-

THE STORY OF A PIECE OF CHALK. 181

pressed into a solid form. As soon as this took place, they passed into real *chalk*, of which I found myself a part. That this flint was originally soft you may see by its having exactly the same kind of shells, &c., sticking in it as you find in the softer chalk. I am reminded of the way in which the siliceous material would separate from the limey mud by a process

Fig. 125.

Microscopic Section of *Polypothecia*, a fossil siliceous sponge, magnified 150 diameters, from Green-sand of Warminster.

which goes on in the manufacture of pottery. When the ground flints have been reduced to a fine powder, and then mixed with clay in the soft putty-like condition, there is a tendency for the silica to separate from the rest, and run together into nodules, so that it is very necessary to prevent by constant agitation. This exactly illustrates how the original silica diffused through the original chalky ooze segregated into flint nodules.

I have a distinct recollection of the creatures that inhabited the sea whilst I was lying along the bottom. I am told there are few objects like them living in the seas of the present day. Even those which approach nearest in resemblance differ in some point or another. The most remarkable of these inhabitants of an extinct ocean were a series of large sponges, called by scientific men *Paramoudræ*, but better known in Norfolk (where I come from) as "Pot Stones." These were originally sponges which grew one within the other, like so many packed drinking-glasses, sometimes to the height of six or seven feet. Through the whole set, however, there was a connecting hollow, which is now filled with hard chalk, the rest being all pure flint. It is very remarkable how these sponges became transformed into their flinty condition. As sponges, they were full of what are called *spiculæ*—that is, flinty, needle-shaped crystals, which act the part of *vertebræ* to the sponge. You may find them in the sponges of the present day. When the "pot stones" existed in this state, as the sponges died and began to decompose, they served as nuclei to all the flinty particles of animalculic shells diffused through the mud, whilst the decomposing animal matter of the sponges precipitated the soluble silica round them, and thus solidified them all together. These replaced the decaying matter of the sponge little by little,

THE STORY OF A PIECE OF CHALK. 183

until the original *Paramoudræ* were turned into "pot stones." That the flint of these "pot stones" was originally soft may be proved by the fact, that fossil shells are often found embedded in it. The other creatures I most distinctly remember are now found in a solid state in the chalk, and are commonly known as "Fairy loaves" (*Ananchytes*) and "hearts" (*Micraster*). They belong to an extensive family still living, and known to the fishermen (who often dredge them from the bottom of the present sea) as "Sea-urchins," on account of their spiny

Fig. 126.

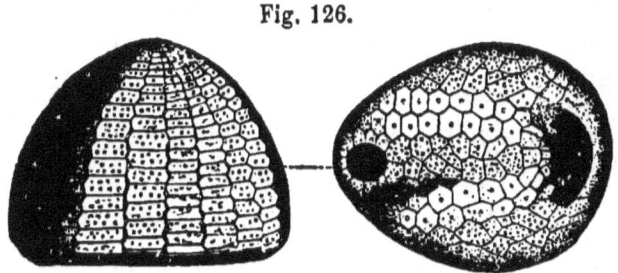

Ananchytes ovata, or "Fairy-loaves."

covering. The existing sea-urchins crawl along the bottom by means of innumerable suckers. Many a time have the fossil fairy loaves thus crept over where I lay. The "hearts" were similarly covered with movable spines or bristles. The family to which these objects belonged is now known as *Echinodermata*, or "spiny-skinned," as the name means. It is as characteristic of the chalk formation as the *Ammonite* family is of the Lias, or the *Trilobites* of the Silurian. Some of these "sea-urchins" were most lovely objects. One group (*Cidaris*) is ornamented by rows of alternately small and large knobs, to which club-shaped spines were formerly attached. These spines are abundant in chalk.

Fig. 127.

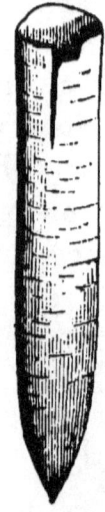

Belemnites mucronatus.

Fig. 128.

Natural Flint cast of *phragmocone* of ditto.

But the commonest objects I remember are those now often found in the chalk as well as the flint, and which are known as "Thunder-bolts." These fossils, however, are individually only part of the creature to which they originally belonged. They were the solid and terminal bones of a species of "cuttle-fish." After the latter had died, and lay embedded in the chalky mud, the soft and fleshy parts decomposed, and left only the harder portions to be preserved. Sometimes the *thorns*, which were attached to the long arms of these creatures, as well as the horny portion of the beak, are also found fossilized. During my time, the *Belemnites* (as these fossils are now called) swarmed the seas in millions; in fact, they were thorough scavengers, and devoured any garbage they came across—dead fish, rotting fairy loaves, &c., and even one another. Here and there, grouped in the hollows of the sea-bottom, lay nests of shells (*Terebratula* and *Rhynchonella*). They are commonly called "cockles," a generic term which fossil shells are always known by to those who

THE STORY OF A PIECE OF CHALK. 185

have not made geology a study. Real *cockles*, however, had not then come into existence. There were a great many species of shells, and these abounded in every sheltered spot. Some few of the fishes were still covered with little enamel plates, instead of horny scales, although it is to this period that the introduction of two of the commonest groups of fishes now living (the *Ctenoid* and *Cycloid*) may be assigned. Sharks also abounded in considerable

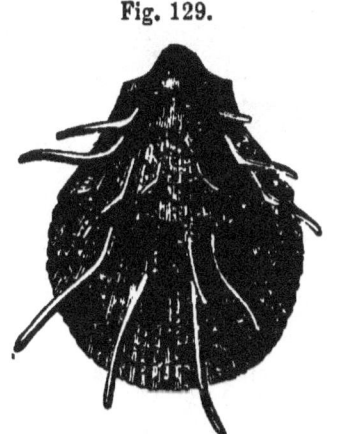

Fig. 129.

Spondylus spinosa—a common fossil in the chalk.

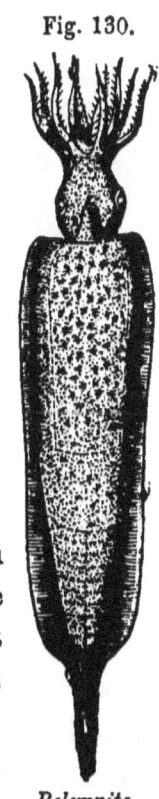

Fig. 130.

Belemnite (restored).

numbers, and I have frequently been witness of the great havoc they made among the shoals of smaller fish. But by far the most gigantic sea-monster was a great marine lizard (*Mososaurus*, or *Leidon*), above twenty feet long, which had teeth implanted in its jaws like bayonets. I have seen its dark shadow pass where I lay, and have beheld the fishes, and even the otherwise bold sharks, dart away in fear. With one or two strokes of its formidable *paddles* (for it had these instead of fins), it could glide through the water with lightning speed. But even this terrible

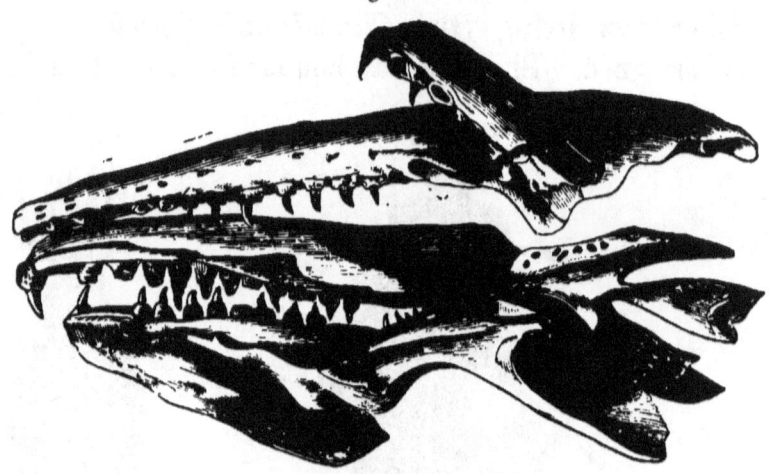

Fossil Skull of the *Mososaurus*.

creature had to succumb to death, and its rotten carcass sunk among the oozy chalk, and there fell to pieces, and became fossilized.

I have not mentioned the fact that before the period when this thick sheet of chalky mud began to form, the Cretaceous epoch had witnessed the deposition of prior formations, all of which are now included by geologists in the same group. The oldest of these is the WEALDEN, a series of fresh-water deposits over a thousand feet thick, wherein you find the monstrous bones of immense land reptiles,

THE STORY OF A PIECE OF CHALK. 187

such as the *Megalosaurus* or "Great Saurian," the *Iguanodon*, the *Hylaeosaurus* or "Forest Saurian," and a host of others. The land plants found fossilized testify to luxuriant conditions of growth. In fact, these Wealden beds were formed at the mouth of a large river, as its Delta. And just as Egypt now owes her agricultural richness to the "spoils of the Nile," so are the hop-gardens of Kent more or less referable to the accumulated spoils of an old continent represented in these Wealden deposits. After the Wealden, there probably elapsed a period of time (for I am now speaking simply of what I have been told) at present unrepresented by any known formations. Then came the LOWER GREEN SAND, or NEOCOMIAN beds, which, in England, give evidence of marine conditions, not only in the fossils, but also in the separate grains of "green sand," which are now known to be in a great measure internal, siliceous casts of microscopic shells. The GAULT succeeds to this sub-division. It is a formation of blue clay,

Fig. 132.

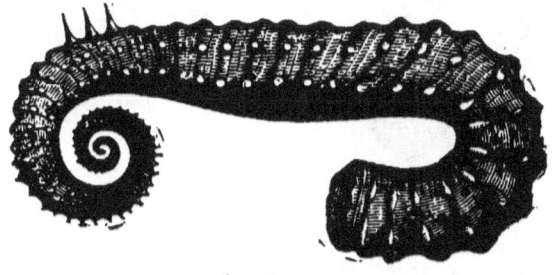

Ancyloceras.

exceedingly rich in modified forms of the *Ammonite*

family, such as *Hamites, Scaphites, Baculites, Turrilites,* &c.—simply chambered shells arranged in various exterior forms. So well have these been preserved in the clay, that they still retain their ancient nacreous, prismatic colours! The UPPER GREEN SAND completes the Lower Cretaceous series, and is chiefly remarkable in England for the quantities of so-

Fig. 133.

Scaphites—a Lower Cretaceous fossil.

Fig. 134.

Turrilites.

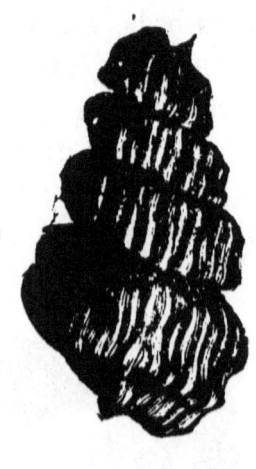

Fig. 135.

Portion of *Turrilites nodosus.*

called "coprolites" it contains. These are simply shells, bones, &c., changed into phosphate of lime by some subsequent chemical process.

Time would fail me to tell of *all* the creatures which lived in my native sea. I remember that, after long ages had passed away, tremors were again and again felt to shake the sea-bottom. It was evident that some earthquake action was at work over a considerable area. By-and-by, we found the water getting shallower, and that the light came through the waves more clearly. The sea-bottom was being upraised; and at length what had formerly been ocean, became an extended mud-flat. The sea was drained off, and covered land which had sunk as ours had risen; and thus the two changed places. The upheaval went on, and the chalk hardened into its present solid state, and became a land surface.

Do not imagine that this upheaval was a sudden and violent process, as some have thought; on the contrary, it was exceedingly slow. The exact spot where I was born was at hundreds of yards depth of sea-water, and the upheaving process was probably not greater than at the rate of a few feet a century. From this you may form some idea of the time it took to lift me from my briny bed to the fresh air and hot sunshine. Meantime, whilst the chalk formation, of which I was an infinitesimal portion, was thus being upheaved, the sea was at work in other localities depositing strata similarly to the manner in which I had been originated. Not a single moment was idled away. The forces of Nature know no Sabbath—they must toil on from the creation to the final consummation of all things! The great work of the sea, ever since the waters were divided from the dry land, has been to lay the foundations of future continents, and even of mountain-chains. Her own barriers have thus been erected by herself, and then as slowly frittered away in order to establish them elsewhere. Geologically speaking, a "new earth" is always being formed! The old one is gradually altered, particle by particle, just as the human body changes its physiological structure, and yet retains its own individuality.

When I did appear above the surface of the sea, it was to form part of an extensive chalky mud-flat. Far as the eye could see, this monotonous landscape stretched away. Here and there an arm of the sea extended, as if old Neptune were loath to quit his sway and to see his recent territory possessed by his rival Tellus. The pasty mud hardened on the surface in the hot sunshine (for the latitude of what is now Great Britain then enjoyed a sub-tropical climate). The upheaval still proceeded, until, at length, after century upon century had passed away, the solid chalk was lifted high enough above the waves to form a tolerably steep coast-line.

For a long time, the hardened *new-born* chalk was perfectly bare. There was neither soil nor vegetation upon it. It extended in an undulating area, just as the sea-currents had carved it, for hundreds of miles. Wind and rain at length formed a light, chalky mould, which was rendered somewhat sandy by the admixture of flints that had been broken up and pounded into dust. Sea-birds lived on the adjoining sea, and for centuries the chalk surface served them as a refuge from the storm, and to build their nests upon. Their excrements, together with the light mould I have spoken of, laid the first foundations of the soils and subsoils which covered me up. Some of the birds left undigested seeds, brought from other lands, and these took root and flourished. The wind came laden with minute spores of moss and fern, and soon thick brakes and morasses clothed the marshy places with cheerful green. An occasional palm-nut was stranded upon the beach, where it grew, and shortly afterwards bore fruit, that spread itself in huge palm forests over an area which, many centuries before, had been nothing but an extensive and barren chalk-flat. In this manner a sub-tropical vegetation covered up the chalk of which I formed part. It has not taken me long to tell, in a general way, of the changes which were thus wrought, but it required hundreds of thousands of years to produce them. After the upheaval had continued for a long time, it suddenly ceased, and the chalky continent, with its wealth of virgin forests and innumerable inhabitants, remained at rest, and in this way the great TERTIARY epoch was ushered in! The ordinary physical laws of nature were in operation, just as they are now. I ought to have told you that the chalk continent extended from the west of Ireland, through Russia, as far as the coasts of what is now the Mediterranean Sea. It is also more than probable that there was a northerly continuation of land across the Atlantic into America. Existing oceans, seas, lakes, and rivers are the results of subsequent processes, which, as may be imagined, took up an immense period of time to bring about.

CHAPTER XI.

THE STORY OF A LUMP OF CLAY.

"CAIN : And those enormous creatures,
Phantoms, inferior intelligence
(At least so seeming), to the things we have passed,
Resembling somewhat the wild habitants
Of the deep woods of earth, the hugest which
Roar nightly in the forest, but ten-fold
In magnitude and terror; taller than
The cherub-guarded wall of Eden, with
Eyes flashing like the fiery swords which fence them,
And tusks projecting like the trees stripped of
Their bark and branches—what were they?
    LUCIFER :   That which
The mammoth is in thy world;—but these lie
By myriads underneath its surface."
BYRON's Cain.

N outline of the biography of even such a humble individual as myself will not be without interest. I need not introduce myself in learned mineralogical language; for there is not a boy living, old or young, who has not made practical experiments on me. But as clay is not limited to any geological formation, but occurs most abundantly in the later deposits, perhaps it may be as well for me to say to which period I belong.

In the older rocks, what was once clay has since

taken the form of slates or shales, subsequent alterations having brought about this change. I may say, therefore, that I belong to that period termed the *Eocene*—a period remarkable for the great influx of warm-blooded animals. This period is the first of that last great division of geological time called the Tertiary. Of these I shall speak presently.

The "London Clay," as it is termed, is the parent deposit of which I am elected spokesman and representative. London has been chiefly built out of this huge bed of clay; whence its geological name. I have a dark bluish-brown appearance, and in some places the fossils enclosed are assembled in great abundance.

Do not confound me with the clay beds which will be referred to by-and-by, and which belong to the Glacial or "Ice" period. No mistake could be greater, although very frequently our general appearance is much the same. It is when you compare the fossil remains found in our beds only, that you would form a just opinion. I was born ages before the clay above mentioned, and, although of marine origin, I came into the world under vastly different circumstances. When I was born, a nearly *tropical* climate existed in what is now Great Britain—when my neighbour was formed the climature was *arctic*. I made my appearance at the commencement of the Tertiary epoch—he did not come until the final close. Between this beginning and end, this extreme of warm and cold climates, a long period of time had elapsed,

THE STORY OF A LUMP OF CLAY. 195

marked by the deposition of thick strata, some of whose members will presently tell you what occurred meanwhile. But, from the time when I was formed to the present, I know there exists a gradual series of beds, in which fossil plants and animals are imbedded, whose types link those of the past with the present living fauna and flora of the globe.

The *Eocene* formation comprehends other strata than that of which I form a part, but I do not think I am egotistic in stating that ours is regarded usually as the principal member. The total thickness of these beds is over two thousand feet. The upper series are well developed in Hampshire and the Isle of Wight, where they bear evidence of having been deposited in fresh water. These are represented on the Continent by the beds of the Paris basin, famous to geologists as having yielded to Cuvier the first materials for the young science of comparative anatomy.

Fig. 136.

Nummulites.

Section of ditto.

Taking the upper Eocene strata in England, you find a gradual transition from purely marine to purely fresh-water conditions, the Headon series containing shells and other organic remains usually found under both circumstances. The Bracklesham sands are crowded with fossil shells, chiefly of *Turitella*, indicating how slowly such beds must have been formed, and how suitable was the ancient sea-bottom to the luxuriant development of these molluscs. I should also mention that underneath the London clay proper is a series of strata, chiefly of sands and gravels, ranging to a total thickness of nearly two hundred feet. My hearers who have carefully studied the geology of older formations, will see that a marked feature about these newer deposits is their very *local* extension. Whereas the older beds are almost world-wide in their distribu-

Fig. 137.

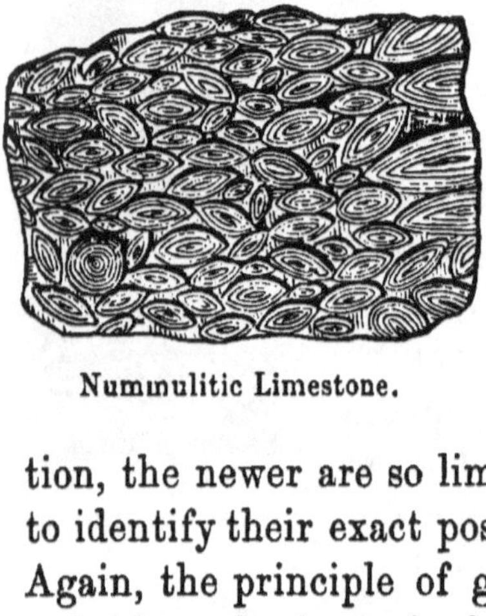

Fig. 138.

Nummulitic Limestone.

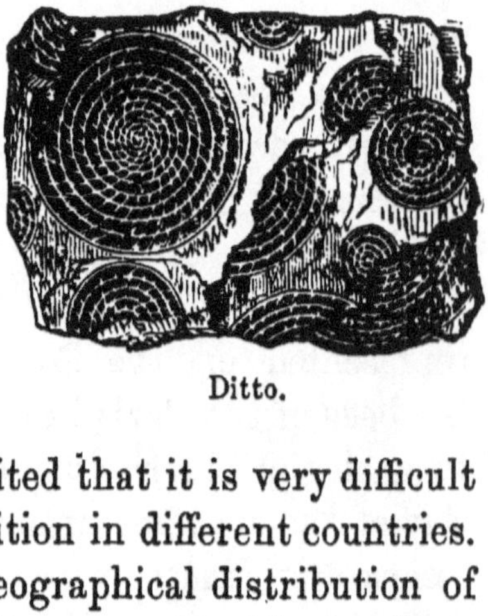

Ditto.

tion, the newer are so limited that it is very difficult to identify their exact position in different countries. Again, the principle of geographical distribution of animals and plants is felt more palpably in these newer than in the more ancient organisms. In the old rocks all over the world you see fossils common to them, but every stratum in the more recent deposits is marked by its own suite of shells, &c., just as every sea now possesses its own peculiar forms of life.

I was formed along the bottom of the sea, at no great distance from land, and yet far enough off for the sediment brought down by the rivers to have had its coarser particles precipitated before it reached the area over which my parent stratum was laid. Consequently, the muddy matter which there fell to the bottom was of a very impalpable character. The distant land was watered by large rivers, whose mouths debouched into the sea, and furnished it with the sedimentary material whose accumulation, to the thickness of nearly five hundred feet, ultimately formed the London clay. This land was clothed with a gorgeous and luxuriant flora, more like that fringing the banks of the Indian rivers, or the islands of the Malayan Archipelago, than any elsewhere growing in the world. Principal among the tropical forms were the palm-trees, whose graceful leaves hung over the water, and were reflected in its rippling depths. The succulent fruits of these palms fell in the stream in immense numbers, sometimes literally covering the surface, and were carried seawards. In some places where the clay was forming, these fruits, now known as *Nipadites*, accumulated to an extraordinary thickness, as in the Isle of Sheppy, where no fewer than a dozen species have been met with. In this locality alone, no fewer than one hundred and six species of plants have been found. You will see the correctness of my inference that an Indian climate and scenery existed in England during Eocene times by-and-by;

but, meantime, I may say that the only places where palms now grow, whose fruit nearest resembles these of the London clay, are the Moluccas. Tree-ferns and fan-palms, also, were not lacking in the brilliant landscape; whilst *Anonas*, or "custard-apples," gourds, melons, bread-fruit trees, &c., completed the list. The rivers which ran through these thickets of tropical vegetation were haunted by crocodiles and gavials, lying in wait to seize the harmless *Palæotheria* which might come to drink or to bathe themselves in the stream, after the fashion of their nearest living representatives, the tapirs. Opossums swarmed in the forest, and there is good evidence for believing that, towards the close of the period I am describing, monkeys were introduced in what were then English woods! At dusk, large bats, not unlike those of the Indian islands, made their appearance. Some of the fish which still lived in the rivers were *ganoids*, that is to say, had bony-plated enamelled scales, like the Polypterus of the South African rivers. The remains of these fishes and bats have been found in some abundance near Woodbridge, in Suffolk. Lazily lurking in the flowery brakes of the forest were huge serpents, some of them as big as the boa-constrictor, and possessing characters now distributed among that class, and the pythons, colubers, &c. In the rivers, and also in the adjacent seas, swam terrible water-snakes, of an enormous size, and with vertically flattened tails, the better to enable them to swim.

THE STORY OF A LUMP OF CLAY. 199

As you would expect from such an association of aquatic dangers, many of the land animals fell a prey, and portions of their carcasses were either deposited in the river mud or carried out seawards. Hence I can tell you something of them, and point out a few leading peculiarities. Chief and commonest among them were the tapiroid animals, to which I have already alluded. These harmless creatures were lighter built than the modern tapir,

Fig. 139.

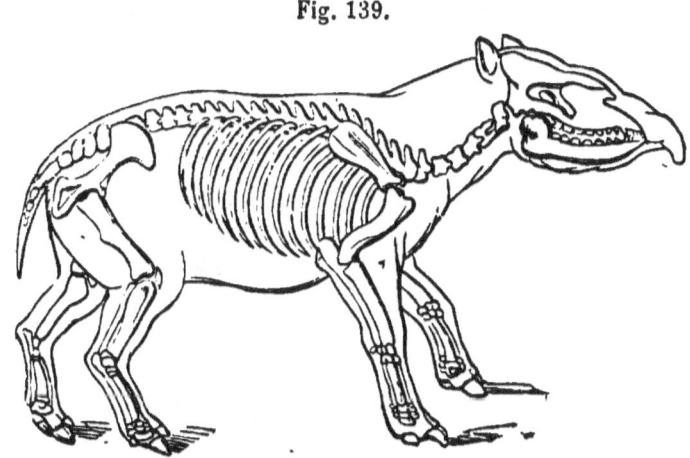

Palæotherium magnum.

although, like that species, they had a short proboscis. Their name of *Palæotherium*, or "ancient beast," is well deserved, as, with the exception of the marsupials, or pouched animals, they are really the oldest warm-blooded quadrupeds with which I am acquainted. They were thick-skinned or "pachydermatous" animals; but, like many of the early types, possessed characters which are now more or less distributed among at least three different groups. The modifications of the higher animals, at the time I am treating on, were necessarily fewer than at present, when such an enormous zoological and physiological "division of labour" has ended in more marked specific specialization. Hence the *Palæotheria* had characters which relate them to the tapir, horse, and rhinoceros! About half a score different species lived together, their sizes ranging from that of a decent horse to that of a pig. Closely

Fig. 140.

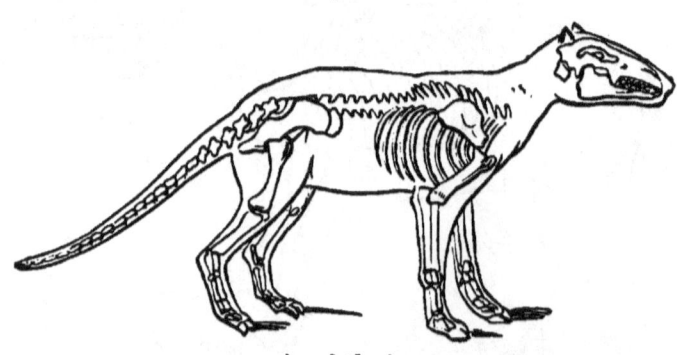

Anoplotherium.

allied to this extinct creature was the *Anoplotherium*, or "harmless beast," as both its name and its structure implied. The most remarkable feature about this creature was its long and powerful tail, which helped it when swimming, just as that of the otter does now. The *Anoplotheria*, however, were perhaps more abundant in what was then France than in England. Some of them were very small, not much larger than a rabbit, whilst the largest certainly did not stand higher than three or four feet. They usually frequented the marshy places, and were very fond of wallowing in the mud. Like their relatives first mentioned, they had various zoological peculiarities, among which was the additional relation to the modern camel. The *Chæropotamus*, or "river hog," was also a genus of the thick-skinned tribe, and stood really as a link between the *Anoplotherium*, and the modern *Peccary*. Its habits, however, were not so harmless, as its teeth indicate a tendency to carnivorous habits. The *Dichobune*—so called from the deeply-cleft nature of its teeth—was allied to the group I am describing. The *Hyænodon* was a truly carnivorous animal, its jaws being even better adapted for cutting flesh than those of the modern feline tribe. In some parts of Europe there abounded an animal called *Anthracotherium*, from its remains occurring in the peat-bogs or lignite-beds of this age. Like that just described, it was of flesh-eating habits; as was also another, very nearly allied to the modern weasel. I have not time to notice the birds and insects of this period; suffice it to say, that the latter included forms now to be met with only in tropical districts. But I hope I have been successful in showing the peculiarities about the terrestrial animals, and you will have no difficulty in seeing how important these extinct types are to the naturalist, in enabling him the better to fill up his natural history plan. These "missing links" thus connect groups of living animals which otherwise would never have been harmoniously blended. It is the moral of Mirza's vision over again—the extinct forms have fallen through the trap-holes of the great viaduct of life, whilst only the recent forms have arrived safely at the other side!

You will have seen that, as far as it goes, the testimony of the mammalia is supplementary to that of the vegetation, &c., all tending to prove what I first stated—that a nearly tropical climate ruled in English latitudes during the Eocene period! The evidence of the marine organisms (with which, of course, I am better acquainted) is exactly to the same point. Just as the Tertiary epoch is remarkable for its large introduction of higher types of animal life, so it is also for the greater influx of genera, animal and vegetable, of living types. For the first time, among shell-fish, you recognise in the fossils of these deposits, forms which are common in existing seas. But it is not in British latitudes, but in *tropical*, that you meet with living genera allied to the fossil. The old *Nautilus* still kept its place, and several species lived in English seas, although it is now scantily represented only in the Indian Ocean. Huge *Volutes*, beautiful *Cones, Mitres, Terebella, Rostellaria, Typhis*, &c., abounded; and the very mention of these names at once conveys to the mind of the conchologist ideas of tropical seas. The fish which lived in the same seas were also of a type commoner to warmer areas than to ours. Many species of sharks abounded, some, as for instance, *Carcharodon*, being of immense size; for the teeth of the largest have been found six inches long, and five broad at the base. Turtles lived in these seas and bred there; for carapaces of all sizes, from the juvenile to the adult, are deposited in that part of the mass to which I belong, forming the Essex cliffs. As you are well aware, the turtles are now almost entirely confined to the tropical and sub-tropical districts.

You see, therefore, that I have abundant evidence for warranting me in my statement, that at the time I was born, a warm climate prevailed here. What it was before, I cannot say; but I know that even before the close of the Eocene period, this warmth had already decreased very considerably. You will, of course, remember that between the beginning and close of this period, there had elapsed time sufficiently long to enable more than two thousand feet of material to accumulate as strata. The changes which took place in the physical geography meantime were very great. I am speaking of a time when those high mountains, the Alps and Pyrenees, had not been elevated—nay, when the rocky material now forming a portion of their flanks was being deposited along the sea-floor!

In England and France, marine conditions had gradually given place to lacustrine, and large lakes had occupied the area previously covered by the sea. During the time that these changes were going on, the climature was slowly toning down. The fossil vegetation met with very abundantly in strata of Upper Eocene age in Hampshire, show you this very plainly. Although it includes types now peculiar to warmer regions, it is not so plainly tropical. The succeeding age, the Miocene, bears out what I say, and from the period of my birth until the present, the register of the climature is very faithfully kept in the strata of the earth.

Fig. 141.—Ideal Landscape of the Miocene Period.

CHAPTER XII.

THE STORY OF A PIECE OF LIGNITE.

"Sweet was the scene! apart the cedars stood,
A sunny islet open'd in the wood
With vernal tints the wild-brier thicket grows,
For here the desert flourished as the rose;
From sapling trees with lucid foliage crown'd,
Gay lights and shadows twinkled on the ground;
Up the tall stems luxuriant creepers run
To hang their silver blossoms in the sun;
Deep velvet verdure clad the turf beneath,
Where trodden flowers their richest odours breathe;
O'er all, the bees with murmuring music flew
From bell to bell, to sip the treasured dew;
Whilst insect myriads, in their solar gleams,
Glanced to and fro, like intermingling beams;
So fresh, so pure, the woods, the sky, the air,
It seem'd a place where angels might repair,
And tune their harps beneath those tranquil shades,
To morning songs, or moonlight serenades."
JAMES MONTGOMERY.

PERSONALLY, I do not think I am such a familiar object, in England at least, as some of my fellow story-tellers. In some parts of Germany and Switzerland, and even in Devonshire, I am much better known under the name of "brown coal." The name I have assumed at the head of this story indicates, although under a Latin form, my vegetable origin.

Of my affinity to the common household coal I will speak presently. My appearance bears out my Latin name, for few would mistake my mineralized woody structure for anything else than it is. Notwithstanding my dull brownish look, and the general absence of that pitchy glossiness which characterizes true coal, I have been formed under very similar conditions to the latter. My history is not less romantic—nay, in my belief, is even additionally so, on account of my having come into existence at a comparatively recent period, geologically speaking. The epoch of my birth is distinguished by the appearance of many genera of animals and plants which are still in existence. These, it will be seen presently, by their occurrence in other parts of the world besides Europe, indicate the immense amount of physical changes which have caused them to take up geographical stations so far away from those in which they were evidently first created.

The epoch of my birth was briefly referred to by the last speaker. It was the *Miocene* period, during which Europe was dotted by great lakes of fresh water, and covered with a flora more magnificent than any she had been clad with since the world began. The scanty species of the Carboniferous period pale before the gorgeous varieties of the Miocene. The flora extended to the very North Pole itself! I am speaking of a period when no ice-cap existed in Arctic regions; but when Iceland, Spitzbergen, and Greenland were clad with evergreen

THE STORY OF A PIECE OF LIGNITE. 209

shrubs; of a time when the Old World and the New were connected by an extension of land of which the Japanese islands, the Aleutian islands, and Vancouver's Island are now the only existing outliers. Central Europe alone maintained no fewer than three thousand known species of plants! Of these, eight hundred species of true *flower-bearing* plants, besides ferns, &c., are found fossilized in the strata called the "Molasse." There is not the slightest

Fig. 142.

Deinotherium (restored).

doubt that this number did not form half the real number, for these two thousand species have been entombed and fossilized wholly through accidental causes.

The temperature of this period was considerably higher than it is at present, although not near so elevated as in the previous *Eocene* epoch. The nature of the plants found fossilized indicates an elevation of about sixteen degrees above what it is now. Hence

P

with physical circumstances suitable, one cannot wonder that a luxuriant vegetation covered every available spot of the dry land. As to the causes of this increased temperature, and, still more, of the extension of the Miocene forests to the very North Pole itself, I can only speculate. It is generally thought, however, that they were due to astronomical conditions of the northern hemisphere, partly similar to those now feebly affecting the southern, and also to such an arrangement of physical geography as insured the highest degree of heat and genial moisture. But even these conditions will not account for plants to which *light* is such a necessary stimulant, growing within the Arctic circle, where there is a continued darkness for months together. I must give it up, seeing that eminent scientific men are in a quandary about it. All that I can say is, that no *geographical* agencies alone will account for the physical circumstances of the Miocene epoch.

The Miocene strata, as I think I have before remarked, are most interesting to biologists, inasmuch as it is here that they meet with the most abundant evidences of the direct ancestry of living animals and plants, which, since then, have been distributed by subsequent physical changes over the surface of the existing dry land.

The fossil plants found in the lignite beds where I lay, before I was disinterred by the curious geologist, to tell him my personal experience, themselves assist me in unfolding a wondrous tale.

Lignite beds of Miocene age are to be found in Germany, Switzerland, Italy, Austria, Scotland, Ireland, Devonshire, Iceland, Spitzbergen, Greenland, Vancouver's Island, the Alaska islands, and elsewhere. All the plants forming this *lignite* afford most indisputable proof of their having grown on or near the spots where they are now met with. The petals, stamens, and pistils of the flowering plants are preserved in the fossil state, together with even the pollen! Then you have the seeds, in various degrees of ripeness, whilst the leaves of many of the fossil plants have also fossil *fungi* on their backs, just as living plants are troubled with "smut," "bunt," or "rust" now.

The ferns are to be met with in the circinate or crosier-like condition, as well as with the ripe spore-cases, ready to burst, on the backs of their fronds. Nothing could be more conclusive as to these various plants, flowering and cryptogamous, having grown near where they are now found in a fossil condition. The facts I have mentioned will show you they could not have been drifted to their present high latitudes by any *flood* or *deluge*, for that would most assuredly have disturbed such minute evidences of local growth as every bed of lignite affords.

Taking this fossil flora in its general character, you will find that it is not so much what you would call *European* as it is cosmopolitan. Of the eight hundred species of flowering plants which geologists have already discovered in the lignites of Switzerland, no fewer than three hundred and twenty-seven species are *evergreens*. Now, at the present time, evergreens are considered peculiar to climates where ahe winter is mild, and therefore where the leaves are not often shed, as Italy, for instance. The majority of the species found fossil in Switzerland and in Germany have, since the Miocene period, migrated to the southern states of North America. The next percentage continued European. Then, in succession, you find other species which have since been transferred to Asia, Africa, and even to Australia. The preponderance of the *American* types, both of plants and insects, is the peculiar character of the Miocene fossils in all the deposits of the Old World. That I was perfectly correct in my statement about the general increased temperature of this period will be evident when I submit to you a few analytical facts connected with this fossil flora. You will have to seek for the European types by the shores of the Mediterranean, and for the Asiatic in the Caucasus and Asia Minor, generally. The *camphor*-trees—now such a characteristic element in Japanese scenery—are very abundant in the fossil condition in Miocene strata as far north as Iceland, Spitzbergen, and Greenland. How imposing was the vegetable kingdom in Central Europe at this time, you may guess by my enumerating a few of the commoner genera.

The *Smilax* grew everywhere, only equalled in

THE STORY OF A PIECE OF LIGNITE. 213

abundance by the *Dryandroides*. Nine species of *Fig-trees* are known, whose nearest analogues now flourish in India, Africa, and America. The *Proteacea* family was very abundant. *Fan-palms* were a peculiar feature in the Miocene landscape, together with occasional *Flabellarias*. Other species of *Palm* were not lacking to adorn the scenery with their graceful foliage. Then we had abundance of *Tulip-trees*, *Magnolias, Banksias, Sequoias, Vines*, &c. You may guess, therefore, at the lovely aspect of the Swiss, Italian, German, and English lakes, set in a frame of such lovely vegetable forms, and whose banks were haunted by animals (which I shall presently describe) whose forms and affinities were quite as foreign to anything existing in Europe as can possibly be imagined.

I was exhumed from my silent position in the pretty valley of Bovey Tracey, in Devonshire, where lignite occurs in several seams. There is not that abundance of vegetable forms stored up here as is to be met with elsewhere, especially in Switzerland. As far as I can remember, only about fifty species of plants are known from this English deposit. The intervening beds tell a tale as to the denudation of Dartmoor, and how the overlying beds came to be chipped off the hard granitic boss. Twenty of the plants found fossilized in this my birthplace are common to those met with under similar circumstances in Switzerland. They are principally *Evergreen Oaks, Fig-trees, Vines, Laurels, Gardenias*, &c.

Miocene beds are met with also in the Isle of Mull, and at Antrim, in Ireland, where the basaltic columns of your Giant's Causeway are of this geological age.* The floral yield of these beds, however, has been very small compared with the same strata elsewhere. A peculiar species of *Fern* grew in what is now the Isle of Mull; but which was, at the time I am speaking of, part of an extended connection with Ireland. The greatest interest connected with these beds is, that they contain evidence of *the last active volcanoes* in the British islands.

The Greenland lignite beds have yielded many hundred species of fossil plants; but their character is hardly so well known as those of other deposits, although it tells the same tale of a mixed flora. The Iceland strata contain no fewer than four hundred and twenty-six species of true flower-bearing plants, exclusive of those belonging to the cryptogamous class. Among them you may find such familiar types as the *Willow, Juniper, Rose, Oak, Plane-tree, Maple, Vine, Walnut,* &c., all of them now living further south, either in the Old World or the New. The reason why the southern states of North America are now occupied by a flora which I have shown you was decidedly *European* during the Miocene period, is that it subsequently migrated thither by way of that continuous land, whose

* Many of the basaltic dykes found in the north, as well as such outbursts of igneous rock as those in the neighbourhood of Edinburgh, are of Miocene age.

outliers are to be found in the Aleutian islands. They were driven to their present southerly habitats by the gradual growth of the great Arctic ice-cap during the Pliocene epoch, and which, in extending so far beyond its present limits during the Glacial period, caused temperate plants to take up positions even under the equator; but at sufficient height to find a temperature analogous to that of their northern home. The further you go *east* in the Old World, the more numerous relatively are the living species which occur in the fossil in the Swiss lignites. The *Salisburia* is now limited to the Japanese region, although it is found fossil in the Pliocene deposits of North America. There are more than three hundred existing species of plants common to the Southern United States and Japan, that are not to Europe. So that, in this respect, Japan is more nearly related to the New World than to the Old.

I have gone into this detail because, although the vegetable forms which enter into my composition are so like those now in existence, as to suggest a recent geological period, yet their cosmopolitan distribution from European centres, the subsequent depression of dry land to become sea-beds, and the uplifting of sea-bottoms into dry land, and even to high mountains, all proclaim the great lapse of time which must have ebbed away since then! Many of the great fresh-water lakes I spoke of just now, set in their frame-work of a southern vegetation, had rivers and streams which supplied them with water. The *deltas* of such streams are still visible in many parts of Central France. The boughs of the overhanging trees, and the host of leaves which were shed in the autumn time, thickly strewed the surface: gradually settling to the bottom, they there formed those beds of woody lignite of which I form an insignificant part. In some of the Swiss lakes there were precipitations of limy matter going on, and these enveloped the leaves, &c., with thin films of carbonate of lime, so as to preserve every vein, mid-rib, and ornamental marking. The fish, such as the *Roach*, &c., as well as fresh-water *mussels*, which lived in the lakes, have their remains occasionally found in numbers. In Central France, there are beds of some feet in thickness actually made up of the accumulated tubes of *caddis-worms!* More than a thousand different species of insects have been obtained from the lignite beds of Switzerland, so that you may guess at the lively sounds which animated the old Miocene woods. Gorgeous butterflies, allied to existing Indian forms, slowly flapped their way through the bosky thickets. Hosts of white ants, or *Termites*, of at least ten species, built their earthy mounds; myriads of insects, of various orders, dropped into these extinct Swiss lakes, poisoned by the mephitic gases which were sometimes evolved in great volume. Among the fossil insects you may recognise forms which mankind now consider *pests*, although they have an antiquity so much greater than themselves. These include the Dung-beetles, Lady-birds, Earwigs, Glow-worms, Dragon-flies, Honey-bees, &c.

Fig. 143.

Skeleton of *Mastodon giganteum*.

I must say a few words respecting the creatures which lived in these magnificent primeval forests. Troops of *monkeys* were not wanting, of which the remains of at least *three* different genera are known. The *Dryopithecus*, or "Tree-ape," lived in France. It was arboreal in its habits, and in stature was equal to that of a man—in fact, it was even a more man-like ape than any now existing. In Greece there lived a genus called *Semnopithecus*, and in the forests where the Pyrenees now rise was another, named *Pliopithecus*. Huge tigers (*Machairodus*) haunted the thickets, scaring the light antelopes and deer. Along with the tree-monkeys were species of *Opossum*, not much unlike those now living on the same trees in the United States. Huge *Deinotheria* frequented the marshy swamps—creatures with downward-bent tusks, and, in natural history position, perhaps intermediate between the Tapir and Elephant families. The *Mastodon* was the characteristic and commonest type of the elephants, noticeable chiefly for its straighter tusks, and more particularly for the mammilated shape of its huge teeth, which, however, were only employed on vegetable diet. The rivers swarmed with many species of *river* or *wart hogs*, associated with *Hippopotami*, *Tapirs*, &c. Herds of wild *oxen* roamed over the plains, their weaklier members falling a prey to the huge tigers, bears, and hyænas which had appeared on the stage of creation by this time. The Deer family had also come into existence, and abounded in great numbers. What was said of the mammalia of the Eocene period—viz., that some of these species combined characters which are now distributed among three or four, is more or less true of many of the Miocene animals. I have mentioned the *Deinotheria* as instances. The *Hipparion*, or three-toed horse—very numerous at this time—was another, inasmuch as it possessed affinities with the ruminantia. In the Miocene deposits of the Sewalik Hills, in India, the "missing links" are even more numerous: chief among the forms there to be met with, is the *Sivatherium*, a huge *four-horned* deer, which connected the ruminant family with the pachyderms. It had a long snout, or proboscis, like the elephant, which creature it nearly equalled in size and bulk. But the most remarkable animal which then lived in India was a huge *Tortoise*, now extinct, whose entire length was over eighteen feet, breadth eight feet, and height over seven feet! I doubt whether the whole records of geology can bring forth a reptile more peculiar, or built on a huger scale, than this. Associated with it are the remains of several species of crocodiles, which then, as now, lived in Indian rivers. The *Giraffe, Camel,* &c., were then Indian mammals, although they are now limited to Africa. In North America you may find other strata of Miocene age, as in Virginia, Nebraska, &c. Most remarkable there are the fossil remains of animals which afterwards became locally extinct; as, for instance, the Horse, Ox, &c. These active creatures swarmed over American plains at the time I am speaking of, just as the Bisons and Wild Horse now do further south. But the latter have thus run wild since their introduction by the Spaniards, whereas during the Miocene period they were natives of the New World, and lived on the same areas with Mastodons, Hippopotami, and Elephants.

You will have seen that the peculiarity I mentioned earlier in my story, as to the chief feature of the Miocene flora being its extended geographical distribution since it grew so luxuriantly in Europe, applies almost equally to the animals. It seems so strange to imagine native horses and elephants in America, and native monkeys and tapirs in England! But I am speaking of facts about which there can be no possibility of mistake. I have only briefly glanced at the chief vital features of this interesting epoch, but my hearers will admit the world was then anything but a desert, although its most highly-endowed tenant—that which then occupied the place now maintained by man—was only *a long-armed monkey!*

The familiarity of the animal and vegetable types of the Miocene epoch, and their great resemblance to, if not identity with, species now living, will cause you to think that it was not so far removed in time as it really was. It is only when your attention is drawn to the physical changes which have gone on since then, that you grasp the idea of unlimited time more fully. Great mountains have been upheaved from the sea-bottom, and continents depressed to form sea-beds, since the events occurred which I have been describing. It was a period when volcanoes were active in Great Britain, and when, in Central France, they threw up great cones of ashes, lava, and scoria, nearly equal in height to Vesuvius or Etna! The Alps, Pyrenees, Himalayas Andes, and other great mountain-chains were then either not elevated at all, or much below their present loftiness. The area of the Swiss fresh-water lakes and of the dense Miocene forests became gradually depressed, until it was a sea-bed, tenanted by hosts of marine mollusca, fish, cetaceans, &c. This great change took place even within Miocene times, for the marine deposits just mentioned belong to the uppermost division of the formation. I cannot speak of the great changes which subsequently swept over the northen hemisphere, of the formation of the great Arctic ice-cap, which spread over temperate latitudes, and drove animals and plants as from another violated Eden, this way and that, until they ultimately occupied their present habitats! All this is matter of fact, as well as matter of geological history; but I cannot be supposed to remember everything that took place since I was born!

Fig. 144.

Skeleton of the *Mylodon*, a fossil American Sloth of Pliocene age.

CHAPTER XIII.

THE STORY OF THE "CRAGS."

"Fens, marshes, bog, and heath all intervene;
Here pits of *Crag*, with spongy, splashy base,
To some enrich th' uncultivated space:
For there are blossoms rare, and curious rush,
The gale's rich balm, and sun-dew's crimson blush."

CRABBE'S *Borough*.

IT may be that some of the friends who are good enough to listen to what we have to say, do not understand what is meant by the term "Crag." Some of our fellow story-tellers have already remarked, that many of the terms used in geology have been borrowed from vulgar use and elevated into scientific expressions. It is necessary to understand the latter before much progress can be made. This, however, is not the place for explaining any other than our own. "Crag," then, is a word common in the Eastern counties, especially in Norfolk and Suffolk, and is applied to those thick beds of marine shells whose history we purpose relating. Ask any person living near the localities where these strata crop out, and they will soon tell you, in a dialect you will find it difficult to understand, to what the word is applied.

Geologically, the "Crag" beds belong to that period of time known as the *Pliocene*. They are deeply interesting, on account of their connecting the past with the present. They also give you a good idea of the physical and climatal conditions of this country just before the extreme and lasting cold of the Glacial epoch set in, and testify to general circumstances not greatly unlike those which now prevail in these latitudes. We "Crags" are three in number, of which the oldest is that known as the *Coralline*. Then comes the *Red*, and lastly, the *Norwich*. The former goes also by the name of Older Pliocene.

We must take you back to a period—that of our birth—when the climate was rather warmer and milder than it now is. A good portion of Suffolk was then lying under a tolerably deep sea, along whose floor beds of shells were forming. The genial temperature of the water was favourable to the development of animal life. Hosts of beautiful echinites (*Temnéchinus*) slowly warped themselves over the smooth bottom. These creatures subsequently became extinct in English areas, and naturalists believed they had passed out of existence altogether. We hear, however, they have been met with quite recently whilst dredging in deepish water off the coasts of Florida, on the other side the Atlantic. You may guess, therefore, the time which has passed away since the Coralline Crag was formed, by the agencies which have slowly driven a once English inhabitant to take up its isolated abode in American waters. The mollusca literally swarmed over the Suffolk area, and it is out of their broken and disunited shells that we "Crag" beds have been formed. Chief among the generic forms were the *Astartes*, whose specific abundance was only excelled by their individual powers of multiplication. Next came the *Pectunculus*, whose members literally swarmed. The *Cyprina* was not absent, and its beautiful valves are among the chief spoils to be obtained at Orford, in Suffolk. One genus, *Cardita*, is also largely represented, and you may frequently disinter it with both valves still united. No fewer than three hundred and fifty species of mollusca lived in the waters of the Coralline Crag sea; and in the beautiful cream-coloured deposits you may pick these out with as much ease as you would the empty valves on some sea-beach. To those who are fond of conchology, and who love still more to read of the simple but profound lessons which fossil shells teach, we would recommend a visit to those parts of Suffolk where we lie in our original repose. It is like walking over the dried-up bed of a recently-existing sea, and obtaining those secrets which the dredge and other instruments can so imperfectly explain in these days. Besides the great number of species of mollusca found here, and in addition to the Echinodermata, or "sea-urchins," there are no fewer than one hundred and thirty species of *Bryozoa*, or "sea-mats," which have been discovered. Some of them, such as *Fascicularia*, are quite unlike anything now existing, although they lived in what were then British waters, at a period so geologically recent. Corals, all of them belonging to the solitary kinds, are also plentiful, and their beautiful shapes are only excelled by the ornate sculpturing of the "sea-urchins." Altogether, therefore, you may form some idea of the rich treat for the naturalist which is to be obtained simply by "collecting" in our beds; whilst, if your philosophy goes deeper, you will not be long before you come to some such conclusions as the following, all of which form a veritable portion of our life-history.

The sea of the Coralline Crag was subject to occasional extremes. On its floor were met species of shell-fish which are now regarded as indicating wide differences of climature. The Astartes decidedly point to northern conditions; but such forms as *Pyrula, Voluta,* and *Cassidaria,* as distinctly point to warm waters. We can hardly speak positively on this point, but we think these extremes may have been produced by alternate currents of warm and cold water, which, as we have heard, are found to exist in the deeper parts of temperate seas at the present time. Whether or not, it is certain that such circumstances would only make the life-forms more various and the species more abundant. The total number of shells which you may call "southern," met with in this the oldest Crag, is twenty-eight—not a large number, you will say, out of the total. But, small as this number is, it will assist us in explaining to you the gradual change of the physical conditions which occurred during the Pliocene epoch. Some of them were driven away from these latitudes by the encroaching cold, and step by step migrated southerly. One species, doubtless the lineal descendant of those which lived over what is now called Suffolk, is met with in the West Indian seas. Most of the "southerly" shells, however, are to be found in the Mediterranean.

By the slow accumulation of dead shells, corals, &c., cemented by the smaller tests of foraminifera, the Coralline Crag eventually attained a thickness of fifty feet. It was slowly upheaved, and subjected to great erosion by the action of marine currents, which scooped out great hollows. When the upward movement was arrested for a time, in these

THE STORY OF THE "CRAGS." 227

Fig. 145.   Fig. 146.   Fig. 147.

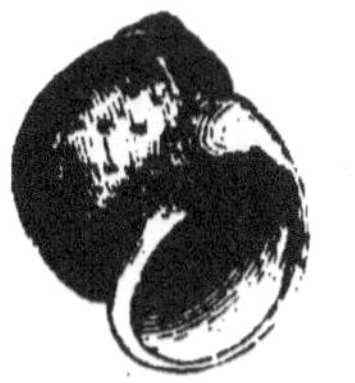

Nassa reticulata.   *Pectunculus glycimeris.*   *Natica catenoides.*

Fig. 148.   Fig. 149.   Fig. 150.

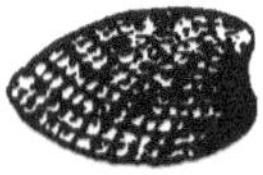

Emarginula reticulata.   *Trochus ziziphinus.*   *Cyprea Europea.*

Fig. 151.   Fig. 152.

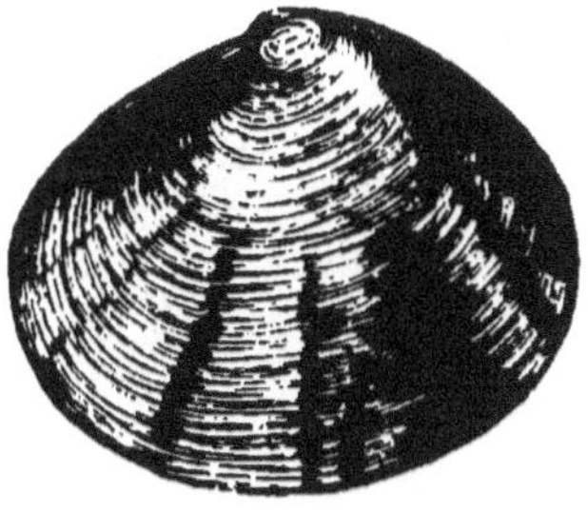

Pecten varius.   *Tellina crassa.*

COMMON RECENT SHELLS OF THE RED CRAG.

hollows was thrown down another and later series of deposits, termed the "Red Crag." This Crag, whose prevailing colour gives to it its name, has a much wider extension than the older member of our series. Just before it was formed, the same wear-and-tear which had so effectually cut down the Coralline Crag also denuded the underlying London Clay. For ages before the depression took place which brought the Crag seas over Suffolk, this had been a land surface, over which had roamed hosts of wild and extinct animals. The wear-and-tear had loosened and washed out the fossils of the London Clay, so that underneath the Red Crag, and with the latter resting on it, you find a bed of stones in which are huge teeth of sharks, bones of whales, teeth of tapir, elephant, mastodon, &c. The stones are those so-called "coprolites" which make the Red Crag so valuable. These are nothing more or less than phosphatic nodules or lumps, in a re-deposited state.

In this Red Crag sea there lived over two hundred and fifty species of shell-fish, among which, however, you will only find about thirteen of the "Southern" forms. The "Northern" types are also on the increase, so that you have in these two facts an indication of an increased rigour of climate. The sea was not so deep as during the formation of the older crag, so that you get a great many more shallow-water shells, among which those of the Limpet family are most abundant. The small single corals were very numerous in places, and the little cowrie-shells literally swarmed everywhere. That the water was shallow you may see for yourself whenever you visit a Red Crag pit, for you cannot fail to be struck by the lines of false current bedding which everywhere meet your eye. The rough marine action, testified to by these phenomena, ground up the more delicate shells into the branlike appearance of which the matrix of the crag is composed.

Extending in a north-easterly direction, towards the conclusion of the Red Crag era, and when its beds had been formed to a depth of at least twenty feet, was a shallow estuary, which ran sinuously through the bare chalk into what is now Norfolk. It occupied the very site of the city of Norwich, and reached its head about four miles beyond, where a small river poured its waters into it, so as to produce brackish water conditions. You will see, therefore, that this later, or "Norwich Crag," as it is usually called, was merely a fluvio-marine extension of the more purely marine Red Crag. Owing to its being formed under different conditions, the fossils of the Norwich Crag differ very much from those of its older brethren. You meet with no corals or other shells which indicate tolerably deep water. Instead you have abundance of periwinkles, cockles, mussels, whelks, purple shells, &c., associated with myraids of *Tellina* and *Mactra*, as well as wentle-traps and *Cerithia*.

Associated with these are brackish-water shells, and such purely fresh-water mollusca as *Lymnea*, *Planorbis*, &c., and even land-snails, which had been brought down by the tributary streams, and eventually strewn along the bottom of the estuary where the Norwich Crag was slowly forming. Altogether, no fewer than one hundred and twenty species of mollusca have been derived from this bed. Underneath it you may see a similar stone bed to that underlying the Red Crag in Suffolk, and, like it, testifying to its having been an old land-surface of the solid chalk; for here are abundant remains of deer, elephant, rhinoceros, mastodon, &c., where bones were strewn over it long before it was depressed to form the bottom of an estuary.

Such are the relative geological conditions of us three Crags. After the formation of the last, a depression ensued, which brought the sea over what had previously been merely an estuary, and along its floor was formed another bed of crag, in which marine shells only have been met with. At Aldeby, on the borders of Suffolk, you may see the shells of this bed occupying their original position, the *Myas*, for example, being found erect in the sand. Neither in the old Norwich Crag, nor in this later bed, do you come across any "Southern" shells; whilst it is evident that the percentage of "Northern" species was proportionately increasing. This is good evidence of the fast-encroaching cold—a cold which shortly afterwards set in, as the drift-beds overlying these crags, and into which the uppermost beds silently pass, plainly attest. The Upper Crag, indeed, is a sort of bracket between the Pliocene and the Pleistocene, or "Glacial" series.

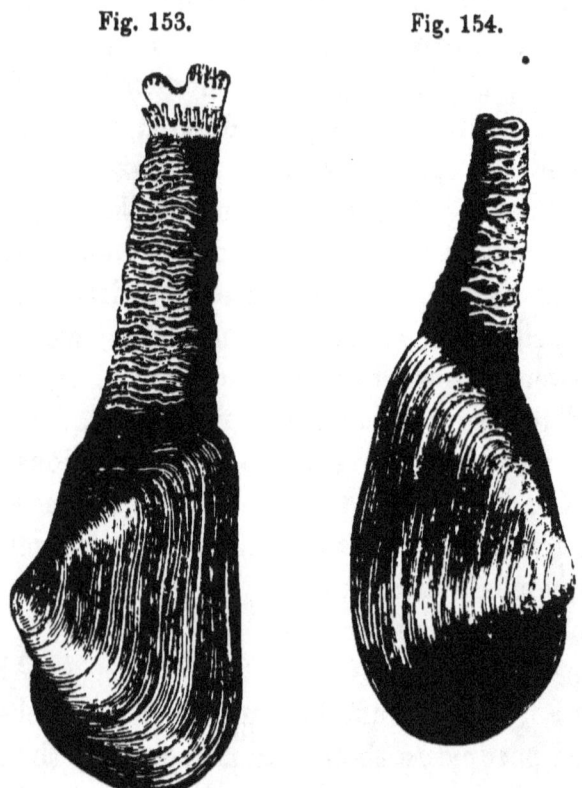

Fig. 153.   Fig. 154.

Mya truncata, in living position.   *Mya arenaria*, in living position.
COMMON RECENT SHELLS OF THE NORWICH CRAG.

It is interesting to note, as you analyze the shells of the crags we have mentioned, how the percentages of recent or living shells to those which are extinct bear out their relative ages. Thus you find less than ten per cent. of *extinct* shells in the Upper Crag just named. In the Norwich beds there are eighteen per cent., in the Red Crag twenty-five, and in the Coralline Crag thirty-one. How long it is since the Norwich Crag was formed, you may gather by the fact that some of its representative shells are now living only in certain parts of the Pacific!

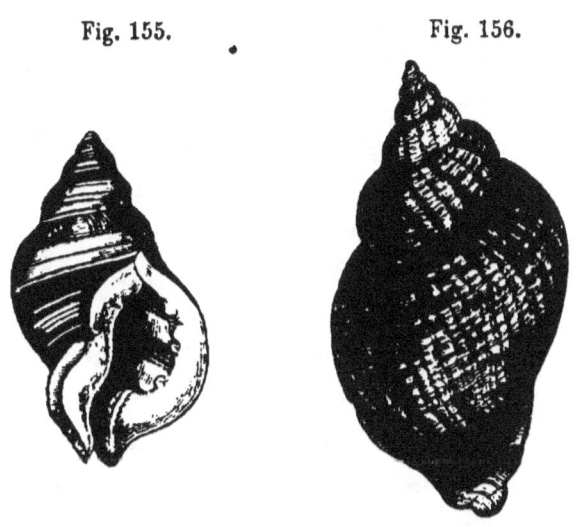

Fig. 155.   Fig. 156.

Purpura lapillus.   White Whelk (*Buccinum undatum*).
COMMON RECENT SHELLS OF THE NORWICH CRAG.

Along those parts of the Norfolk and Suffolk coasts most dreaded by our gallant seamen, where the storms of the German Ocean expend their greatest fury, and where the solid land is consequently being gradually washed away, you may see one of the most interesting geological phenomena in Great Britain. Cropping out from beneath the steep cliffs at Cromer, Mundesley, Happisburgh, and South Wold, and extending along and forming the very floor of the existing German Ocean, is an old Forest-bed. It has been traced more or less for a distance of nearly fifty miles, although its actual landward and seaward extensions will perhaps never be known. Here you may plainly perceive the indurated soil whereon trees grew, much resembling those now living in British latitudes, with only one exception—the Norway Spruce Pine. This, it is true, is now naturalized amongst us; but I am speaking of a time when it was an indigenous English tree. The soil of this Forest-bed contains ample evidence of the flora which formerly grew upon it, and of the many strange creatures which then sought the shelter of the greenwood. The age of the Forest-bed is very great, and seems to be all but contemporaneous with that when the Norwich Crag was formed. Since this forest was in its primitive condition, the present overlaying cliffs of clay and sand have been formed under glacial conditions, as an extensive marine sheet of mud! After some north-west gale has been blowing a few days, the rising sea will have stripped the sand off the beach, and you may then behold stump after stump, stool after stool, of these ancient forest trees imbedded where they grew, their roots still spreading through what was once rich soil, but which is now sea-bottom. Not unfrequently the bones of elephant and rhinoceros may be seen sticking out of the denuded cliffs like the bones out of a pigeon pie! The leaves, branches, and fruits of Oak, Willow, Hazel, Elm, Alder, Pine, &c., as well as the fronds and rhizomes of ferns, may be easily obtained. Not only do we thus get glimpses of a quiet English forest, before the glacial winter set in here, but also plain evidence of fresh-water lakes, along whose floor the rich black soil was first formed. The surfaces of these waters were adorned with white and yellow Water-lilies, Buck-bean, Duck-weed, &c., all of whose remains are met with. Out at sea, the trawling-boats are always bringing up in their nets patches of the semi-hardened soil of the Forest-bed, as well as bones, tusks, and teeth of its elephants, &c., thus indicating its great seaward extension. The lines of your great poet, Tennyson—

"Now rolls the deep where grew the tree,"

are nowhere in Great Britain more certainly realized than in this Pliocene deposit.

But if the trees and plants are familiar forms, the remains of the animals which then haunted the forest and morass seem strikingly un-English. At least two species of elephant—perhaps three—trumpeted in these pre-adamite woods. No fewer than seven or eight species of deer roamed about. A huge beaver (*Trogontherium*) inhabited the lakes and marshes; the great Cave Tiger preyed on the herbivorous tribes. One elephant (*Elephas meridionalis*) must have been at least sixteen feet high, according to the size of its leg and thigh bones. Its tusks have been found measuring fourteen feet in length! Rhinoceri, hippopotami, oxen, swine—all were associated in this locality. It was as if an immense menagerie of foreign animals had been let loose in a modern English wood!

There are beds of the same age as the Coralline Crag in Belgium, but these are often hardened into stone. They may, and perhaps are, somewhat older than those we have been attempting to describe, and may be regarded as connecting the Miocene with the Pliocene period, just as the last-mentioned crag-bed connects the latter epoch with the Glacial. We have ample proofs that this Belgian bed extended across the German Ocean into Suffolk, where it was broken up, and the fragments, rolled and angular, are often found in abundance at the base of the Red Crag, and are known by the local name of "Box-stones."

In Sicily, beds of Pliocene age abound, and have been uplifted to three thousand feet above the sea-level since the time when the Norfolk and Suffolk beds were formed. Many of the shells spoken of, which migrated from English latitudes during the later Pliocene, and when the cold was increasing, took up their habitats in Sicilian seas, and are now found fossilized in the limestones. Since then their descendants have returned to their original English home, and, as the oyster and mussel, administer to modern English appetites. The oldest of these Sicilian beds, perhaps contemporaneous with the

Coralline Crag, was strewn over an area that was subject to volcanic shocks. Occasionally volcanic ashes were intercalated with the shell marl. At length, by the simple accumulation of volcanic ashes and lava, during a slow elevation as well, a great mountain eleven thousand feet in height was formed! That mountain was *Etna*, and the Pliocene shell-beds at the height of one thousand two hundred feet along its flanks indicate its recent origin. In Italy, just above Florence, there was a great fresh-water lake, into which the rivers occasionally carried carcasses of mastodons, elephants, &c. The deposits which formed along its bottom accumulated to two hundred and fifty feet in thickness. All over the Northern hemisphere great zoological as well as physical changes occurred during the period of the "Crags." Animal life slowly prepared for that great event which wrapped Europe in glacial ice during a long winter, tens of thousands of years' duration! All these facts may be more or less accurately and minutely read off in the sometimes loose and unconsolidated strata of the Pliocene age, of which we Crag-beds are the English representatives.

CHAPTER XIV.

THE STORY OF A BOULDER.

"As a huge stone is sometimes seen to lie,
Couched on the bald top of an eminence,
Wonder to all who do the same espy,
By what means it could thither come, and whence—
So that it seems a thing endued with sense,
Like a sea-beast crawled forth, that on a shelf
Of rock or sand reposeth—there to sun itself."
WORDSWORTH.

FEW of my fellow story-tellers can boast of adventures equal to mine. My life has been a restless one, and to see me quietly reposing in some bed of clay, the non-geologist would little suspect what strange romances I could tell him. I will do my best to recount them. Not many years ago this would have been totally impossible. At that time geology was chiefly made up of guesses, many of which, however, proved to be shrewdly true. The great sheets of sand, gravel, and clay which extend, more or less, over the northern, midland, and eastern counties of England—as well as over the Continent and in the United States of America, were supposed to have been the *débris* left by Noah's Flood, and were therefore called "Diluvium." But facts (stubborn things!) have accumulated in such numbers that it is now totally impossible to hold such an idea—much as many people may wish it. It is seen that the period of time when such beds were formed was as peculiar as those of other formations, and that the physical circumstances, if not the peculiar life-forms, marked it off distinctly from the rest. Hence the name now given to it of "Northern Drift," or that other of the "Glacial period," which latter I hold to be the most appropriate.

The chief interest of the "Glacial epoch" is the way with which its facts connect older tertiary life-forms and geography with existing species and circumstances. The geologist is able to perceive there was no break, such was originally supposed, but that the present epoch is intimately related to all that have gone before, and is, in fact, a continuation of many of their circumstances. It therefore links the present with the past, in a way for which knowledge-seekers cannot be too thankful. Who would imagine the scattered, disunited beds of clay, or gravel, or sand, could have been so fruitful in geological and even general interest?

Some of my companions may boast of an origin quite the opposite to my own. Theirs deals with intense heat, mine with almost as extreme cold. Of course I am speaking of my present existence as a "boulder;" for before I entered that state I formed an insignificant part of a great and continuous rocky stratum. What this rock was composed of, matters little or nothing, for we " Glacial Boulders " have no such clannish feeling as other geological story-tellers. We are composed of all kinds—and the bed of clay in which we have been deposited may be regarded as a sort of lithological Parliament, in which the representatives of every formation are assembled. But allow me, if you please, rapidly to sketch the outlines of the events which transpired before I was ruthlessly wrenched from my original rocky home, and transposed into a boulder.

As many of my hearers are aware, the earlier part of the Tertiary period was, in England and elsewhere, marked by an almost tropical climate. During the Eocene epoch, the seas of our latitude were inhabited by shells and fish of tropical types. The dry land was clothed with tree-ferns, palms, &c., and these gorgeous forests were frequented by huge serpents, strange-looking, tapir-like quadrupeds, and monkeys. The rivers, also, had their alligators and crocodiles. In short, all the types of land, fresh-water, and marine fauna and flora, which now distinguish equatorial regions, existed in England. The rocks of this period are full of proofs of the truth of what I say. Then gradually succeeded the Miocene epoch, during which the climature was less torrid. Even then, the great Arctic ice-cap had not been formed at the pole; for we have abundant evidence that countries situated far north, such as Greenland and Spitzbergen, were covered with vegetable forms nearly allied to those now living in South Carolina,

Japan, the Cape of Good Hope, and Australia. Then succeeded the Pliocene age, whose climate is abundantly indicated by its fine "Crags," as the beds of shells are termed. The oldest of these is called the "Coralline," and there may be found in it no fewer than twenty-seven species of shells, nearly allied to or identical with those now existing in southern latitudes. The "Red Crag" comes next in age, and this tells you by similar evidence that the climate was gradually getting colder, for the number of southern shells had dwindled to thirteen, whilst there has appeared in English latitudes species allied to those now living in northern seas. Finally, the third, or "Norwich Crag," supplements the teachings of its relatives by its total absence of southern shells, and its much greater proportion of Arctic species. Another bed of Crag, situated some height above this, still further corroborates the remarkable fact I have been narrating, for its greater abundance of northern forms is as remarkable as that of the older Norwich Crag over the red. About the same age as the latter bed is a phenomenon, known as the "Forest bed," which crops out from beneath the steep cliffs along the Norfolk and Suffolk coasts. It is the site of an old forest, now forming the floor of the German Ocean, and the imbedded stools of trees, as well as those of land and fresh-water plants, indicate a temperate mildness of climate, similar to that now marking the British islands—or, if anything, a trifle colder, as the presence of the Scotch fir and Norway spruce pine clearly shows.

My hearers cannot but be struck with the gradual refrigeration of climate, from a tropical or subtropical condition, to a temperate one. Meantime, the slow but sure change from warmer to colder physical circumstances clearly prophesied that the next period would probably be marked by the same law. Such proved to be the case. The change of climate indicated by the several periods I have mentioned, culminated in that "Glacial period" during which my birth as a boulder took place.

After the epoch of the "Crags," a gradual subsidence of England, at least as far south as what is now the Thames, slowly took place. Little by little the whole country sunk beneath the sea, in which, with increasing depth, there came increased Arctic cold. The greater part of Scotland—certainly the whole of the Highlands—was covered with land-ice, or sheets of accumulated snow, frozen into ice. The snow-line—which in England is now some thousands of feet above the ocean-level—then was gradually lowered by the greater cold until it was met with as low as it could possibly creep. The hills of North Wales, Cumberland, Lancashire, and other places also had their ice-sheets and glaciers. To what thickness this great ice-mass accumulated, or from what cause, I can form no idea; but if it was anything like what now takes place in Greenland—and I have every reason for asserting that England at the time of which

I am speaking, experienced Greenlandic circumstances, rather than those of any other part of the world—then this sheet of snow or ice possibly grew to be hundreds, if not thousands, of feet in thickness. Such is the case in Greenland at the present time. The fine snow there accumulates on the mountain-tops, and is only got rid of by its freezing into a sheet, which is always moving down to the lowest level. In temperate and tropical climates, rivers carry off the excess of moisture—in Arctic countries this can only be done by the moving ice-sheets, termed "glaciers." The Greenland glaciers debouch into the sea itself. The ice-sheet forms grand sea-cliffs, hundreds of feet high, into whose bases the angry sea eats caverns, until the toppling mass falls over, and floats away as an iceberg. Or the great ice-sheet thrusts itself right into the sea, creeping along its bottom until it comes to water deep enough to buoy up, break off, and float away the extreme end.

You will have no difficulty in perceiving that the immense mechanical force exercised by such glaciers on the solid hard rocks over which they creep must be immense. You can easily conceive how the latter must be ground down and pounded into mud; and also, how fragments would be broken off, frozen into the great icy mass, and slowly carried away. When that portion of the glacier into which some huge fragment has thus been frozen, reaches the sea, it would be broken off, and floated away as an iceberg, carrying the enclosed fragment of rock with it. Away drifts the iceberg, carried by oceanic currents in a southerly direction, until the warmer waters gradually melt it, and then down drops the rock to the bottom of the sea, to rest perhaps thousands of miles away from its parent source.

The friction of a moving glacier elicits just enough heat to melt a portion of the ice, which flows away as water, carrying with it the finer mud or sand set free by attrition. Hence all the water flowing into the sea is turbid with mud, and this mud, as it gradually settles to the sea-bottom, is there forming what will some day be a geological deposit. In this mud Arctic mollusca live and die, and will also some day be found fossilized. It was in a similar bed to this that I was dropped. Down I sank amid the oozy mud, displacing the strata, and more or less causing them to assume a contorted appearance. Well do I remember the effect produced by the largest boulders, dropped in a similar way into the same strata. They sank so deeply as to cause thin beds of shells, which had previously been horizontal, to wrap over and become almost vertical. In the Norfolk cliffs near Cromer, where what is known as the "Coast Boulder Clay" attains a great thickness, you may see masses of chalk imbedded, which cannot be less

Fig. 157.

Tellina Balthica—a common marine shell in the Boulder Clays.

than two hundred feet in length. The soft sand and clay beds near are so contorted that you would imagine an earthquake had produced the disturbance; but it was caused simply by the melting icebergs dropping their stony burdens. For ages this process went on—the land glaciers grinding down the solid rocks, and the sea currents strewing the *débris* over the ocean floor. The icebergs, also, added no little to the accumulating mass.

I am told that along the North Atlantic sea-floor there is going on a similar deposit. The thousands of icebergs which set out from the north every year gradually melt as they near the more southerly latitudes. There is a great stream of warm water called the "Gulf Stream," which sets out from the Gulf of Mexico, crosses the Atlantic, and impinges on the southern and south-western coasts of Great Britain. When the northern icebergs come into contact with it, they rapidly melt, so that, of course, the sea-bottom in that place might be expected to be heaped up with the *débris* they had dropped. Actual soundings prove this to be the case; so that if the North Atlantic sea-floor could be upheaved, you would have a series of loose deposits of sand, mud, boulders, &c., not unlike those which were formed during my own lifetime.

I am not left without a natural barometer to fix the depth to which the dry land went down. In North Wales is a hill called Moel Tryfaen, and near its summit, at seventeen hundred feet above the sea-level, is an old *sea-beach*, formed when the submergence had reached its maximum. After this there came as gradual an upheaval, and this is marked in various places in Great Britain by a graduated series of raised beaches, ranging in height from that above given to those only a few feet above high-water mark. Gradually the land appeared more extensively above the water. The climate was still intensely cold and Arctic. The icebergs coming from Scandinavia frequently brought with them Arctic plants growing on the frozen mass of gravel or sand. Whenever these icebergs stranded on the coast, these plants were able to migrate inland, and very soon they covered the new land with an Arctic and sub-Arctic flora. Those soft beds of sand or mud lying along the sea-bottom which first came within the influence of the surface-currents, were very much worn away or denuded. This was especially the case with an extensive sheet known as the "Chalky Boulder Clay," from its containing so many small rounded pebbles of chalk, as well as large boulders of other rocks.

Among farmers, this goes by the name of "Heavy lands," and the bed is usually found occupying the highest grounds, having been denuded by marine currents into what are now valleys. A good deal of the material thus worn away was carried by the waves to form beds of later date, which sometimes go by the name of "Post-glacial," although they were really deposited during the Glacial epoch.

THE STORY OF A BOULDER.

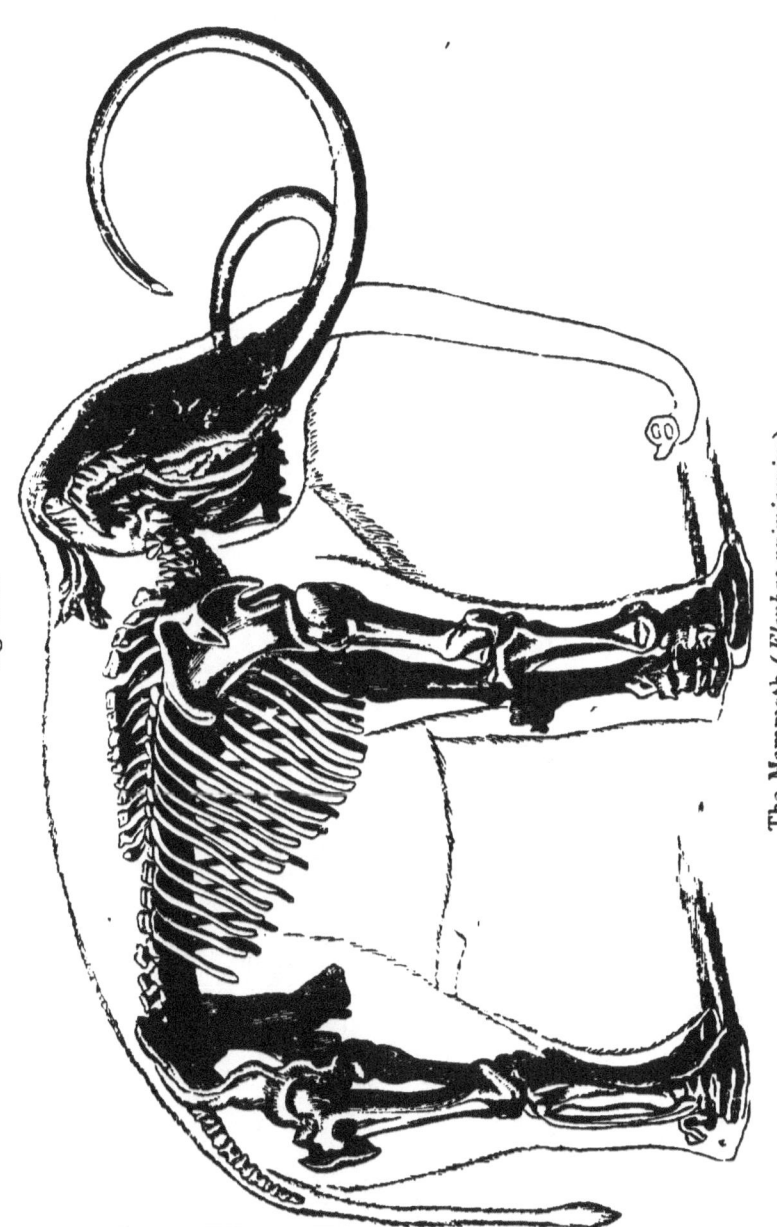

Fig. 158. — The Mammoth (*Elephas primigenius*).

In fact, some geologists are doubtful whether the glacial period is in reality quite over! Of course, we boulders had no means by which we could be transported, and so we were exposed to current-action. The waves rubbed us together, toning down our sharp angles, and very frequently obliterating the scratches and groovings which we had before borne as evidence of our ice-conveyance. In this way a huge plateau gravel or boulder bed was often formed on the highest grounds, the soft matrix having been washed away.

When England was again joined to the Continent, and before the Straits of Dover had been cut out, the European land animals migrated hither. The climate, though still rigorous, was nothing like so cold as it had been during the middle of the Glacial period. Among the animals thus roaming amid semi-Arctic woods and wilds, were the "Mammoth" (*Elephas primigenius*) and the Hairy Rhinoceros. Both these animals were covered with long woolly hair to protect them from the severe cold. Ireland was then joined to England by way of the Isle of Man, and over this extensive prolongation of Europe in a westerly direction, another animal, the "Irish Elk," roamed in great numbers. The Reindeer, Glutton, Lemming, Muskdeer, and other animals affecting high latitudes, then abounded in England, their bones being frequently found in the later deposits, as well as in the cave breccias. An almost Arctic flora covered the plains, and crept up the

250   THE STORY OF A BOULDER.

hill-sides as far as the then perpetual snow-line. Glaciers still debouched through the mountain defiles into the plains, and moraines, or heaps of angular stones, thrust forward by the advancing

Fig. 159.

Irish Elk (*Cervus megaceros*).

foot of the glacier, still remain in Scotland, Cumberland, and Wales, to indicate how far these glaciers travelled. Where the ice-sheet descended from the mountains, of course there was the greatest amount

THE STORY OF A BOULDER. 251

of pressure. Here great hollows were scooped out of the hard, solid rocks, and these hollows are now filled with fresh water, and form the lakes of North Wales, Lancashire, and Cumberland, and, on the Continent, in Switzerland. The Swiss glaciers, by the way, were then much more extensive than they now are. At present their growth is impeded by a warm wind, which accumulates over the Desert of Sahara, in North Africa. But at the time of which

Fig. 160.

Skull and Antlers of the Irish Elk.

I speak the Sahara was a sea, as is indicated by the abundance of ordinary cockles and mussels found a few feet below its surface of terrible drifting sands. Then no warm wind could form, and the European glaciers grew unchecked. Again, the temperate mollusca, such as oysters, cockles, mussels, &c., had migrated from our latitudes, and taken up their abodes in seas which, although farther south, represented in glacial times, as far as the temperature was concerned, the present seas of Great Britain.

As the climate became warmer, the Arctic plants left the lowlands, where they became extinct. Their places were taken by a more southerly flora, which originated in Asia Minor, and covered the greater part of Europe. The Arctic plants occupying the highest grounds, therefore, were the only remains of this once widely-spread Arctic flora, which could find suitable and fitting circumstances amid which they could live. And here the wandering botanist now finds them—living proofs of the truth of what I have been saying respecting the long Arctic winter of the northern hemisphere. Subsequently, Ireland was separated from the Continent, England having been cut off some time before. When the climate had toned down, MAN appeared on the scene. His weapons are found in the most recent of deposits, and his bones beneath the stalagmitic floor of limestone caves. The woolly-haired Elephant and Rhinoceros soon after disappeared for ever; the Glutton, Lemming, Reindeer, &c., like the Arctic plants, migrated with the decreasing cold into northern regions. Meantime, the bottom of the glacial sea had become dry land. The old, hard, and barren rocks had been thickly strewn with rich subsoils, the very elements necessary for agricultural purposes. Nature had done, by means of her glacier and other action, exactly what the scientific farmer sometimes does when he adds artificial manure to improve his soils. She had ground and pounded all the older rocks to make up a new compound that should possess all their valuable mineral ingredients. In this way only could mankind have been blessed with the necessary elements for the purposes of husbandry. Thus, in comparison with other periods, that when man was introduced was especially favoured.

Fig. 161.—Ideal Landscape of the Quaternary Period.

CHAPTER XV.

THE STORY OF A GRAVEL-PIT.*

"Where Tees, full many a fathom low,
Wears with his rage no common foe;
For pebbly bank, nor sand-bed here,
Nor clay-mound, checks his fierce career,
Condemned to mine a channell'd way,
O'er solid sheets of marble grey."

SCOTT's Rokeby.

AM the last of my race. My brother storytellers have had their day, and ceased to be. Had you questioned me a few years ago, I should have been like Canning's Knife-

---

* We are indebted to Mr. John Evans, F.R.S., for the loan of the

grinder, and had none to tell. Even now my story is not complete. New and revised versions are constantly coming out, although their general truth remains unaltered.

Who among my listeners has not played, when a child, in a sand or gravel-pit? We have them in abundance scattered over the surface of the country. But there are gravel-pits and gravel-pits—a difference without a popular distinction.

Those I particularly represent are usually situated on the banks of some great river-valley. Hence their other geological names of "Valley-gravels" and "River-gravels." Frequently they form terraces flanking the present course of the rivers, and you may identify two of these terraces—a low-lying one and a higher. If you could strip off these banks of gravel, you would find the bare rock beneath, or else some thick sheet of boulder clay, which had been scooped out to make the present river-valley. Banked up against these old denuded surfaces are the gravels, whose excavations are so well known as pits. I am one of them, and I propose to tell you my story, as well as I can recollect it. Although I can hardly define the difference between the gravels to which I belong and those which belong to the Glacial series, generally the Middle Drift, yet the practised eye readily learns to detect that there is a difference.

Flint implements illustrating this chapter. They are from his magnificent work on 'The Ancient Stone Implements, Weapons, and Ornaments of Great Britain.'

The pebbles composing our beds are well rounded, showing they have undergone a tremendous deal of wear-and-tear. They are composed of different kinds of rock, just as you would expect when you know they have been washed out of the boulder clays, or brought down by the river in its passage over the outcrops of successive beds. The flint pebbles have generally an *oily* look about them, and

Fig. 162.

Terraced Hills, Glen Colombkill.

all the pebbles are red and ochreous. Their position along the river-valley, however, is always the best test. Some of these valley-gravels are very thick, whilst others extend as mere banks of local distribution. All of them, however, indicate some degree of antiquity, inasmuch as you will find ancient trees growing on the most recent of these terraces, and,

THE STORY OF A GRAVEL-PIT. 257

Fig. 163.

Terraced Hills of the Burren, as seen from the north of Galway Bay.

here and there, old ruins which stand upon them. In fact, the gravel-pits indicate a gradual rise in the land for them to occupy their present heights above the river-level. The gravels were originally brought down by the ancestor of the present river when it was broader and perhaps more turbulent, at the close of the Glacial epoch, when the climature was more severe than it now is, and the quantity of rain and snow which annually fell much greater, so that the river-valley was subjected to great floods, which brought down the materials of which we are composed.

As the land gradually rose, and the climate became more genial and toned down to its present mildness, the waters of the river shrank in volume, until only the present channel was occupied. But the heights to which we river-gravels rise above the water not only indicate how old we are, but, in some cases, go back as far as the commencement of the original scooping-out of the valley itself.

All this would be very interesting in itself, as geological action connecting the most recent of the great physical changes with those we see in operation around us. But the interest of these valley-gravels is still more enhanced when my listeners understand that it is in them that the *first evidences of Man's appearance on the earth* are met with! All my brother story-tellers have had their say, and many of them have described the commonest of the animals and plants of their time; but not one of them mentioned that mankind was living at the time. It was reserved for so humble and commonplace an object as a Gravel-pit to unfold the most important of all geological discoveries. Men have speculated as to their original ancestors living as far back as the Miocene period, but they have adduced no facts in support. On the contrary, I yield nothing but facts, and those in great abundance. In the gravel-pits you meet with the chipped flint implements, of which you have doubtless already heard. They are imbedded, as stones, along with the other material, having been brought down by the ancient rivers in the same way as pebbles.

But they are undoubtedly of *human* workmanship. This cannot be gainsaid. You see this at once by the flints having been carefully, and in many cases *artistically* chipped down to a cutting edge all round. They are generally spearhead-shaped, and about six to nine inches long. Had they not been connected with the question of the antiquity of Man, you would never have heard a word said about their not being of human manufacture. As it is, in order to steer clear of this disagreeable truth, many have invented all kinds of ingenious hypotheses to account for the flints getting chipped in this regular fashion. But it requires far more faith to believe in these theories than it does in the other commonsense inference.

The most damaging fact to them is the *identity*

Fig. 164.

in pattern of these cut and chipped flints wherever they may be met with. Another important incident is this—the chipped flints are only found in the valley-gravels or in deposits of the same age. If they have been chipped by accident, there is no reason in the world why they should not be found in gravel-pits of much older date.

From the time when primitive Man used these flint weapons for almost every purpose, slaying wild animals with them, cutting down trees and scooping them out for canoes, making holes in the ice with them for fishing purposes—since then you can trace the whole history of offensive and defensive weapons. Antiquaries and geologists call these most ancient of implements *Palæolithic*,—meaning, in Greek, that they are the oldest known; and the age in which they were produced consequently is known by the same name. When Man first appeared, if we are to reason by the remains with which we find those implements associated, the woolly-haired Elephant, or *Mammoth*, and the woolly-haired Rhinoceros were both natives of Great Britain. It is frequently objected that you do not find the *bones* of man associated with these tools; but the reason is not difficult to find.

Let us remember how few of the bones, &c., of the ancient Romans and Saxons are met with in proportion to the number of more enduring ornaments, coins, &c., they left behind them. Then again the valley-gravels lie in the line of greatest drainage toward the river, and, as they are porous, the surface water percolates through them on its way to the lowest level. Any particle of carbonate of lime, whether in the form of bone or not, which was deposited in these gravels, would thus be dissolved away. Hence it is that, although the huge bones of elephants, &c., were undoubtedly buried up in the same gravels, we find few or no traces of them. The commonest of their remains are *teeth* and *tusks*, whose dentine and ivory structure saved them from the gradual destruction to which the frailer members of the skeleton were liable.

Fig. 165.

Flint Arrow-head (Neolithic).

Fortunately, there were other agencies at work during the same period, which were conservative rather than destructive. In the fissures of limestone rocks, where water is percolating, the water is usually charged with carbonate of lime. Every drop of water that evaporates on the surface of the walls of a chasm or natural hollow leaves its contained particle of lime behind. This process is always going on, until there has been left on the walls a great fold or layer of what is called *stalactite*. The water drips on the floor, and there a portion is evaporated, the lime being left behind.

As you may guess, the process is marvellously slow, but the layer thus formed on the floor is called *stalagmite*. It is not difficult to see that anything lying on such a cavern-floor would be incrusted over, and eventually covered up. This is what I call a *conservative* process. Now at the time the valley-gravels were forming, savage man was glad to avail himself of any shelter, and the natural caves and hollows of the earth were anxiously sought after, as they are now by the lowest tribes of mankind elsewhere. To such places as Kent's Cavern, Brixham Cavern, &c., savages resorted, bringing with them the fruits of the chase. Here you may find the bones of animals which had been split open in order to extract the marrow, as well as the flint knives and implements of exactly the same kind as those found in a gravel-pit. Over these there has accumulated a layer of *stalagmite* many feet in thickness; thus carrying you back in time as far as does the

Fig. 166.

Palæolithic Flint Implements from Kent's Cavern, Torquay.

knives deposition and origin of the valley-gravels themselves!

Fig. 167.

Portion of Core, from which flint knives have been chipped; Kent's Cavern.

You see, therefore, that the two most accessible groups of facts both point to the same great fact of the antiquity of Man. Succeeding the *Palæolithic* age is that provisionally known as the "Reindeer period," on account of the large number of the remains of that northern animal which have been found in the bone-caves of the south of France. England and the Continent were then subjected to the periodical migrations of Arctic animals, among which were the Reindeer, Lemming, Glutton, Elk, &c. The flint implements found associated with the remains of these animals in the south of France exhibit a superior skill, indicating that man's nature was to progress, even at that early stage.

Rude attempts at *carving* and *drawing* were also indulged in, as examples in your principal museums will attest. Then succeeded the next stage, known as *Neolithic*, or "Newer Stone age," which is distinguished by the greater variety in shape of the flint implements, and, more particularly, by the fact that they are for the most part ground smooth and to a sharp knife-like cutting edge. These weapons, however, are usually found strewn on the surface, or imbedded only in peat-bogs and the most recent of river-deposits. Whereas the *Palæolithic* types are

THE STORY OF A GRAVEL-PIT. 267

Fig. 168. Fig. 169.

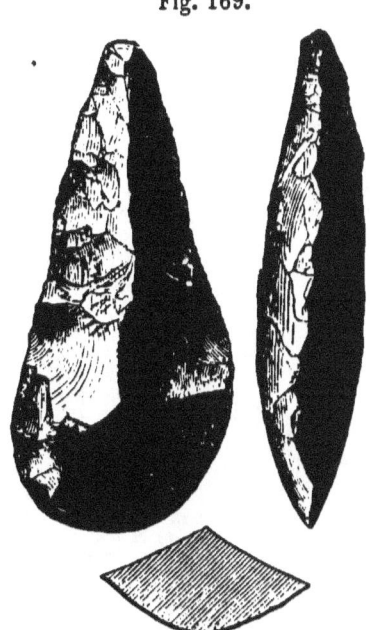

Neolithic Implement, Mildenhall, Suffolk.

Neolithic Implement, Mildenhall— more finished.

Fig. 170.

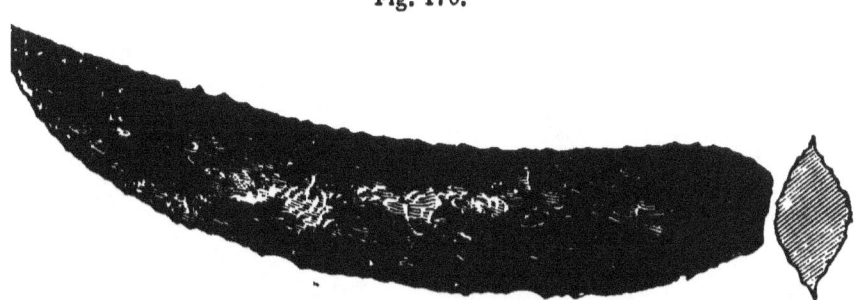

Ancient Neolithic Flint Knife.
NEOLITHIC TYPES OF FLINT INSTRUMENTS.

THE STORY OF A GRAVEL-PIT. 269

limited to valley-gravels and the most ancient of bone-caves, the *Neolithic* show, by their universal distribution and superior workmanship, that they belong to an advanced period. All the savage races still using stone weapons are generally islanders,

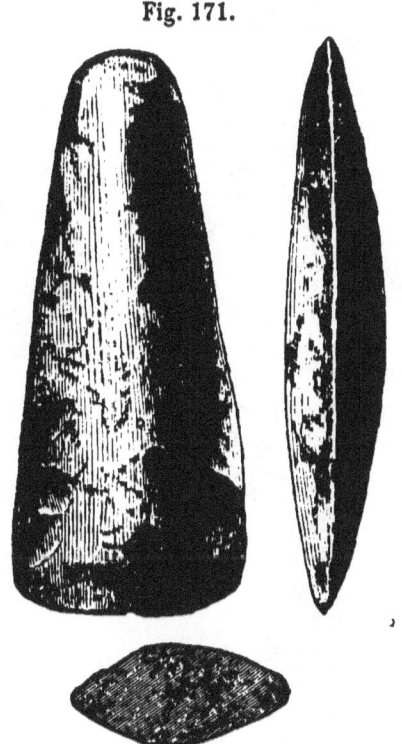

Fig. 171.

Polished Stone Celts (later date), from Cambridgeshire.

cut off from the great centres, so that they are "outliers" of a system once universal. This later period is that of the "Lake Dwellings," which links on to that known to antiquaries as the "Bronze period." To this succeeds the Iron age, and, if you

270 . THE STORY OF A GRAVEL-PIT.

like, the present, or "Steel" age. The two former are historical—come within the range, not only of scientific deduction, but also of written history. I have simply mentioned them to show how, from the

Fig. 172.

Stone Celt (Neolithic), mounted in wooden haft, showing how these implements were used. The haft and weapon preserved in peat, Cumberland.

time when the most ancient and rude of the flint implements were deposited in the river-gravels, there is more or less of an unbroken sequence.

Archæology commences where geology leaves off—the past and the present meet on common ground. Standing on this neutral area, you may gaze backward into the illimitable ages which have gone by, and see the gradual ascension in animal life which began in the dim and distant Laurentian epoch in the animalcule, and has terminated in Man. Looking forward from the same vantage-ground, you may hopefully note the development of society, the growth of civilization, and probability of the unfolding of the social and moral attributes of man as marvellously as the lower animal life has culminated in its existing apex! Throughout, in the buried past, as well as in the yet unfolded future, you never lose sight of the operations of an Almighty Spirit—ever working, never resting!—out of chaos bringing forth order—out of simple protoplasmic material educing the animal and vegetable kingdoms, in all their multitudinous types and varieties, until a small area like the superficies of this planet has teemed with life sufficient to stock a million existing worlds! One generation has passed away, but, in doing so, has furnished a new basis on which the new comer may ascend to a higher physiological platform. Every form, animal and vegetable, has been but the expression of Divine Love, communicating to them the excess of its own joyous life! Every species has been an outwardly crystallized Divine idea. Spirit has clothed itself with matter, until in Man the past and the future have met: the ancient

Greek fable has been more than realized, for it has been true spiritual fire from heaven—given, not stolen—which has been inspired into fleshly clay!

My story is now ended, and, with mine, the series, whose purpose has been to give as plain an outline of the biography of our old world as possible. It will have been seen that a story may be properly read off, even from so common and ordinary an object as a Gravel-pit. In geology, more than any other science, he that humbleth himself shall be exalted! All its objects lie at your feet, and are of the lowliest kind. Not a pebble you accidentally kick before you, not a handful of dust blown by the wind into gutters, not a spadeful of soil turned over, but each is fraught with teaching of the utmost value and of the intensest interest. It is by recognising a *Cause* that you alone can unlock the secret, setting out with the full belief that everything exists by virtue of a right—has resulted, not from accident, but law,—until you arrive at the highest conception of which man is capable,—that the total of these various laws meets and concentrates into one focus, and finds its expression in a personal and Almighty God!

RETROSPECT.

IN the preceding pages we have endeavoured to limn, but in faint and sketchy outlines, the biography of our planet. We now propose still more briefly to connect the scattered ideas into a short summary. Perhaps the most difficult thing a person experiences when he comes into contact with geological teaching for the first time, is the great demand made upon his imagination for the article of *Time*, in which to account for geological phenomena.

It bewilders one to contemplate such a practical eternity, and we ask—"can all this be true?" Many cannot accept the doctrine, but turn away sceptically discontented, thinking they are doing heaven service, by adhering to the older idea that the world is only some six thousand years old, as if the Deity were complimented by supposing His attributes were more honoured by limiting their display to six thousand years, than they are if extended into the past, and made eternal. The more we study the phenomena of geology, however, and the more knowledge of natural science we bring with us to the task, the more profoundly impressed do we become with the vast antiquity of our planet.

Many men fall into the error of supposing that in discovering new laws in the universe, we are exiling the Deity, and giving to the operations of these laws the power that is really His. Even scientific men sometimes speak of the laws of nature as if they were entities, forgetting they use the term simply as a figure of speech. For, as Dugald Stewart has shown, the term "law" can only be applied in its correct sense to *conscious agents*, capable of understanding the rule of conduct laid down, of obeying it or disobeying it. When we apply it to such a system as that which guides the planets, which arranges the animal and vegetable kingdoms, and which directs the operations of physical geography, we are speaking of unconscious objects, which cannot *obey* law, inasmuch as they are not conscious of it. To their relationships, therefore, the word "law" is used as a figure of speech, and limits itself to the mode by which an active Providence is operating on matter.

Though it may seem strange indeed to hear that the world has been in existence millions of years, and that its surface has been covered by numberless creations of animals and plants, yet the true naturalist sees in these extinct faunas and floras, conjoined with the present, only one great and harmonious scheme! In the fossils of the rocks we have a graduated scale of animal and vegetable life, and we have seen how, in spite of the imperfection of the geological record, it is possible to link object by object together, so that, when extended from the remote past to the present, they form a connected chain.

Our planet's earliest existence seems to have been that of a cosmical mass—a sort of world-fog or vapour—something like those revealed by the telescope as being still in existence. Some of the best astronomers have shown that the probable origin of the entire Solar system has been a condensation of this cosmical vapour into planets, satellites, and planetary rings. Whether this was the case or not, it is certain that there is much in the shape and physical constitution of the planets to lend support to the idea. But, with the oblate shape of our globe, and its probable evolution from a cosmical mass, the geologist has little or nothing to do. But he knows, from the fact of igneous rocks having repeatedly been injected into the stratified rocks, so as to bind them together, as mortar does the bricks of a wall, that the interior of the globe still contains molten matter.

The first time the geologist can lay his hand on a formation distinct in its character from the primitive igneous rocks, it is when he comes to the Laurentian system. They are thirty thousand feet in thickness, and so contorted and changed by the pressure, heat, and mechanical forces to which they have been subjected since the infancy of the world, that all original characters have been obliterated. But, by the aid of the microscope, the explorer is yet able to discern that the ancient sea along whose floors these mica-schists, gneisses, and quartzites were deposited as muds and sands was not a lifeless area, but was tenanted by lowly creatures after their kind. The only solitary known fossil from the altered limestones of the Laurentian formation—the *Eozoon*—is sufficient to prove this. And from the occurrence of this lowly-organized creature up to the present, we never afterwards lose sight of the graduated life-scheme recorded in the rocks! There is many a difficult chapter to spell out, many a leaf missing, but there is still sufficient left to interpret the stony scroll.

Above the Laurentian system lies the Cambrian. But we should remember that this classification of the rocks into formations and systems is, at the best, but a harsh and forced one—a remnant of the time, not long ago, when men believed there were distinct creations and destructions of separate faunas and floras. Geological and, in fact, all natural history classification is but an arbitrary arrangement to enable the human mind, in its faintness, to grasp and arrange the multitudinous facts presented to it. In reality there is no separation of geological systems, but more or less of a graduation of one into another. The world's biography is like a man's, not like a butterfly's, consisting of metamorphosed states, each unlike the other, and definitely separated from it. In the Cambrian formation, we find that life, which had begun, as it were, from a point, was radiating like the rays of light from a focus. Here we find the lowest order of shell-fish (*brachiopods*), worms, and, towards the later period, crustacea.

But it is in the Silurian system that we find the stream of life broadening out. The seas are full of coral-reefs, bivalve and univalve shells, huge crustacea, tolerably highly-endowed Trilobites, &c. At the close of the formation, we came on placoid fishes, the first *vertebral* types. Thus we find a *lateral* development of species, in size, and a *vertical* one in organization. Then comes the Devonian, or Old Red Sandstone epoch, whose seas abounded in strangely-clad and gigantic ganoid fishes, and whose deeper waters were busy with the manifold complexities of marine life. The dry land was scantily covered with a thin vegetation of a cryptogamous type, or of the lowliest of the exogens. Great freshwater lakes existed, set in beautiful frameworks of tree-fern and huge club-moss. But it is when the Carboniferous era commences that we find abundant evidence of a dense flora, although one of a very lowly kind. Every foot of dry land, where the circumstances were favourable, seems to have been densely covered with forests, the trees of which now find their nearest allies in our "Horse-Tails" and club-mosses. Enormous *Sigillaria, Lepidodendra,* and tree-ferns constituted this vegetation, whilst there was no lack of species of Conifera. In the Carboniferous limestone period, which immediately preceded that of the coal measures, we have ample evidence of seas in which life was very abundant, where floors were covered with thick submarine forests of sea-lilies, and which had numerous colonies of brachiopodous shells. Cephalopods, such as *Orthoceras*, *Nautilus*, and *Goniatites*, abounded, and thus the huge thickness of limestone rocks grew out of their accumulated remains. The fishes were bony-plated, and, in the structure of their teeth, many of them showed decided reptilian affinities. It was in the waters of the Carboniferous seas that the first reptiles appeared, as the *Archægosaurus*—a creature belonging to the lowest order of reptiles, the amphibia. It exhibits decided affinities to the fish, as the ancient fish do to the reptiles.

In the Permian epoch, geologically brief though it was, the physical geography seems to have been varied. Here we have evidence of a cold climate, and of glacial conditions, during which the "breccias" were formed. Reptiles of a higher class abounded, and these are now known as *Thecodonts*. With the close of the Permian, we have the termination of the Primary, or Palæozoic division of geological time.

The Triassic epoch, or that of the New Red Sandstone, offers to us fresh scenes and new creature forms. Huge frog-like reptiles abounded, and left their numerous foot-prints on the soft muds. In the deeper seas, new species of sea-lilies grew, and new forms of cephalopods, such as *Ammonites* and *Belemnites*, existed side by side with the old-world forms, that were now rapidly dying out. Thus the Triassic limestones of Germany are as crowded with organic remains as the mountain limestone of Derbyshire. Elsewhere, the dry land was covered with saline lakes, or "Dead Seas," along whose floors Rock-salt was deposited. In America, the first *birds* appeared, whilst at the close of the Triassic era in Europe we have distinct and sure proof of the first introduced mammals. The latter belong to the group which all naturalists have by common consent placed at the bottom of the sub-kingdom mammalia. Thus it will be seen that the order in which the new groups of animals appeared on the stage of creation, is also that which we have ourselves arranged, more or less, as that of true succession.

With the Lias, we have the commencement of that "Age of Reptiles" which well deserves the name. New forms of cephalopods appeared, the *Ammonites* literally swarming in the seas, and actually forming limestones by their accumulated remains. New and complete species of sea-lilies grew on the ancient ocean-floors—new plants, cycads and zamias, as well as complex-veined ferns, on the dry land. But the chief animal forms which strike the eye are the reptiles—modified then to every condition of life, as we find the mammalia are now. As *Ichthyosauri* and *Plesiosauri* they were the tyrants of the deep; and as *Pterodactyles* they winged the air like bats, their size being often bigger than that of any existing bird. During the succeeding Oolitic period, huge reptiles lived on land, such as the *Megalosaurus*, *Hylæosaurus*, and *Iguanodon*. Some of the reptiles walked on two legs, like the modern kangaroo, and were decidedly allied to birds. The first known European bird now put in an appearance, its feathers and bones having been found in the Solenhofen slates. It had a long vertebrated tail like that of a lizard, feathered down to its tip. In other respects also it possessed reptilian affinities. Mammalia abounded, but still as marsupials, although there had been a division into herbivorous and carnivorous species. We have evidence of great fresh-water lakes, along whose floors thick beds of limestone were formed by the slow accumulation of *Paludina* and other fresh-water shells. A great river watered a great continent, and at its mouth was formed a Delta, since known as the Wealden formation. Out in the blue sea, coral reefs fringed the rocky coasts; bony-plated fishes and sharks were in plenty, some of the former living on the mollusca.

Then comes that period of great depression when the chalk strata were formed along the floor of a very deep sea, as its organisms plainly prove. For this white chalk is chiefly made up of shells so minute that the naked eye cannot perceive them. Many of the same types of marine creatures still lived, reptiles, brachiopods, and cephalopods. Echinoderms were more abundant than ever, and their remains are to be found in every chalk quarry. A peculiar reptile, the *Mososaurus*, lived in the deep sea, and was a most formidable animal. The sea-bed produced dense crops of sponges, great and small, some of them of as ornate a character as the recent "Venus' Flower-basket." On the dry land, towards the close of the period, there appear for the first known time trees of a higher order, such as the Oak, Walnut, and Elm. Thus came to a close the Secondary or Mesozoic division of geological time, during which we have seen animal and vegetable forms attaining higher and complexer organizations.

The last, or Tertiary epoch, commences with the Eocene beds, in which warm-blooded animals appear so common, that the Tertiary has been not unfitly called "the Age of Mammals." Many of these mammals united characters which since then have been distributed among half-a-dozen later animals. In fact, nearly all the Eocene and Miocene mammals are veritable "Missing Links!" We have, in the former period, evidence of at least a sub-tropical climate in Britain: palm-trees, tree-ferns, &c., grew abundantly. The seas had what we now regard as sub-tropical shells, *Typhis*, *Volutes*, *Cones*, &c., living in them, as well as turtles, sharks, sword-fish, &c. In the rivers, gavials and crocodiles wallowed. Towards the close of the Eocene, *monkeys* made their appearance in English woods; whilst in the Miocene period, they swarmed in several species all over Europe, one of them, singularly enough, being more anthropoid, or "man-like," than any now in existence. Extensive forests of warm temperate plants grew all over the northern hemisphere during the Miocene age; and there does not seem to have then been any ice-cap at the North Pole, for these virgin forests grew in Iceland, Greenland, and Spitzbergen. Elephants and mastodon, camels and giraffes, deer and oxen, now made their appearance. Great fresh-water lakes existed in Switzerland, along whose bottoms the decaying vegetation accumulated to form Lignite beds. In central France, Scotland, and Ireland, volcanoes were very active, as the lava sheets plainly prove.

In studying the Miocene plants, shells, &c., we come across the same genera as are still in existence, so that the naturalist cannot turn away from the impression that many of our modern species are lineal descendants. In the "crags" of Norfolk and Suffolk, this impression rises to a certainty, for in them we actually do meet with hundreds of species of shells of exactly the same kind as those still in existence. These "crag" beds belong to the succeeding Pliocene period, and they tell us very plainly of a refrigeration, or toning down, of the climature. This indication is fulfilled when we study the beds of the Northern Drift—those accumulations of sand, gravel, and clay which occupy the area of the northern hemisphere. These were all formed under glacial conditions, and Europe lay for centuries beneath a thick swathing of land-ice. Arctic plants and Arctic mollusca lived in British latitudes. Our higher mountains sent forth streams of glaciers, which scratched and pounded the solid rocks over which they moved. In British seas icebergs were continually stranding and floating, dropping their burdens of sand and gravel, as well as the huge masses of rock which had been frozen into them. As boulders we frequently meet with these erratics, which had thus been carried miles from their native or parent bed.

It was after the elevation of the glacial sea-bed into dry land, when the climate had toned down from its arctic vigour, although still much colder than it is now, that MAN first appeared on the scene. His rude flint implements have been found abundantly in the valley gravels of existing rivers, formed when those rivers had a greater volume of water than they have now. From that distant time to this, we never lose sight of him and his works, and there is exhibited in his history a similar development, or elevation, from a lower to a higher stage, to that which we have seen marking the lower animal and vegetable kingdoms in their appearance on the platform of existence. But it does not follow that because we can plainly trace the mode in which Deity has chosen to operate, that therefore He has been superseded by His own laws. Rather, it brings Him awfully near, for in the constant regulating and leading upwards of the organic world we never escape His presence!

APPENDIX.

In the foregoing pages frequent reference has been made to geological systems, formations, and divisions, besides the employment of other technical terms. We have therefore given, at the end, the following Table of the British Rocks, from the Catalogue of the School of Mines, Jermyn Street.

THE following TABLE shows the Succession of the *British* Formations, beginning with the newest Strata.

TABLE OF BRITISH FORMATIONS.

CAINOZOIC OR TERTIARY, AND RECENT.

- Recent and Post-Tertiary. . . . .
  - Brown Sand of various ages.
  - Alluvium ,,
  - Peat ,,
  - Raised Beaches ,,
  - Cave Deposits ,,
  - Valley-, or Low-level Gravel.
  - Brick Earth ,,

- PLIOCENE.
  - Newer Pliocene. . . . .
    - High-level Gravel
    - and
    - Glacial Drift } of various ages.
    - (Till and Boulder Clay)
  - Older Pliocene. . . . .
    - Cave Deposits.
    - Norwich Crag.
    - Red Crag.
    - Coralline Crag.

- MIOCENE. . . . .
  - Leaf Bed of Mull.
  - Lignite of Antrim.
  - Bovey Beds.

TABLE OF BRITISH FORMATIONS.

CAINOZOIC OR TERTIARY, AND RECENT—*cont.*	EOCENE.	Upper Eocene.	Hempstead Beds.	Corbula Beds Upper Freshwater and Estuary Marls Middle „ „ Lower „ „
			Bembridge Beds.	Bembridge Marls . . . . „ Limestone . . . .
		Middle Eocene.	Osborne Beds.	St. Helen's Sands . . . . Nettlestone Grits . . . .
			Headon Beds	Upper Headon Beds . . . . . . Middle „ . . . . . . Lower „ . . . . . .
			Bagshot Beds	Upper Bagshot Sand. Middle „ { Barton Clay. Bracklesham Beds. Lower „ Sand and Pipeclay.
		Lower Eocene.	. . . .	London Clay and Bognor Beds. Woolwich and Reading Beds (Plastic Clay). Thanet Sands.
MESOZOIC OR SECONDARY.	CRETACEOUS.	Upper Cretaceous.	Chalk . .	Upper Chalk. Lower „ Chalk Marl. Chloritic Marl.
				Upper Greensand.
				Gault.
		Lower Cretaceous.	Lower Greensand.	Folkestone Beds. Sandgate Beds. Hythe Beds (Kentish Rag). Atherfield Clay.
			Wealden.	Weald Clay.
			Hastings Sand.	Upper Tunbridge Wells Sand. Grinstead Clay. Lower Tunbridge Wells Sand. Wadhurst Clay. Ashdown Sands. Ashburnham Beds.

(Fluvio-Marine.)

TABLE OF BRITISH FORMATIONS.

MESOZOIC or SECONDARY—*cont.*	TRIAS. { OOLITE. {	Upper Oolite. {	Purbeck .	Upper Purbeck Beds. Middle ,, ,, Lower ,, ,,
			Portland .	Portland Stone. ,, Sand. Kimeridge Clay.
		Middle Oolite. {	Coralline Oolite.	Upper Calcareous Grit. Coral Rag. Lower Calcareous Grit.
			Oxford Clay	Oxford Clay and Kellaways Rock.
		Lower Oolite. {	Forest Marble	Cornbrash. Forest Marble and Bradford Clay.
			Great Oolite	Great or Bath Oolite. Stonesfield and Collyweston Slate and Northampton Sands.
			Fuller's Earth.	Upper Fuller's Earth. Fuller's Earth Rock. Lower Fuller's Earth.
			Inferior Oolite.	Ragstone and Clypeus Bed. Upper Freestone. Oolite Marl. Lower Freestone. Pea Grit.
		LIAS. {	Upper Lias {	Lias Sands. ,, Clay and Shale.
			Middle ,,	Marlstone.
			Lower ,,	Lias Clay, Shale, and Limestone.
		Upper Trias. {	Koessen Beds {	White Limestone or Westbury Beds. Bone Bed.
			Keuper . . {	Red Marl and Upper Keuper Sandstone. Lower Keuper Sandstone and Marl (Waterstones

APPENDIX.

TABLE OF BRITISH FORMATIONS.

				England.	Scotland.
MESOZOIC OR SECONDARY—*cont.*	**TRIAS**—*cont.*		.....	Muschelkalk, absent in Britain. St. Cassian Beds, ,, Dolomitic Conglomerate (of Keuper or Bunter age, Somerset and Gloucester).	
		Lower Trias.	Bunter	Upper Red and Mottled Sandstone. Pebble Beds. Lower Red and Mottled Sandstone.	
PALÆOZOIC OR PRIMARY.			Permian	Upper Red Marl. Upper Magnesian Limestone ⎫ Lower Red Marl. . . . ⎬ Zechstein. Lower Magnesian Limestone ⎭ Red Marl, Sandstone, Breccia, and Conglomerate (Röthe-liegende).	
	CARBONIFEROUS SERIES.		Coal Measures.	Upper Coal Measures. Middle ,, Pennant Grit. Lower Coal Measures Gannister Beds.	Upper Coal-Measures.
				Millstone Grit (Farewell Rock).	Moor Rock.
			Carboniferous Limestone.	Upper Limestone Shale (Yoredale Rocks). Carboniferous Limestone. Lower Limestone shale.	Upper Limestones, Edge Coals Series, Lower Limestones. Sandstones, Shales, and Burdie House Limestone.
	OLD RED SANDSTONE AND DEVONIAN.		Devonian	Upper Devonian and Petherwin Limestone. Middle Devonian Limestone and Cornstones. Lower Devonian.	

Table of British Formations.

PALÆOZOIC OR PRIMARY—cont.	SILURIAN.	Upper Silurian.	Ludlow . .	Tilestone. Upper Ludlow. Aymestry Limestone. Lower Ludlow.
			Wenlock .	Wenlock Limestone. Wenlock Shale, Sandstone, and Flags. Woolhope Limestone. Denbighshire Grits, Shales, Slates, and Flags.
			. . . .	Tarannon Shale (Pale Slates). Upper Llandovery Rock. (May Hill Sandstone). (Pentamerus Beds).
		Lower Silurian.		Lower Llandovery Rock. Caradoc or Bala Beds.
			Llandeilo .	Upper Llandeilo Flags and Limestone, &c. Tremadoc Slates.
				Lingula Beds.
		Cambrian.	Cambrian .	Harlech Grits, &c. Purple Slates and Grits (St. David's). Llanberis Grits and Slates. Longmynd Rocks. Red Sandstone and Conglomerate (Scotland).
		Laurentian.	. . . .	Gneiss of the Lewis, &c.

The above table, it will be seen, has reference to what are called "typical" sections of the different formations—that is, to those exposures of rocks which have been most studied and are best known. Hence many of the names of the formations and groups are more or less local, as this was the only available manner of naming the sub-divisions when geology was a young science. It may seem absurd to speak

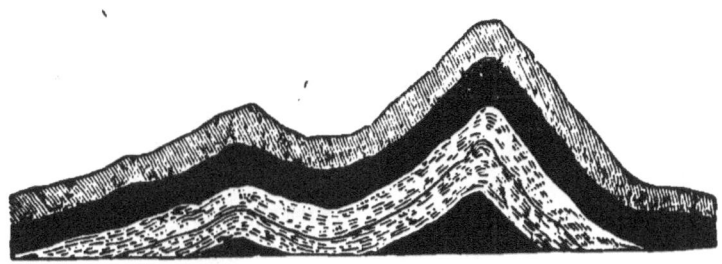

Illustrations of "Anticlinal" strata.

of "Devonian" (or Devonshire) rocks in America, for instance, as we are forced to do by this method of nomenclature; but, after all, it is not more singular than to call the language spoken in the United States *English*.

The following terms are largely used in every work on geology, even the most elementary, and, as they cannot well be avoided, perhaps it will be desirable to give a short explanation of them. STRATUM (from the Latin word meaning

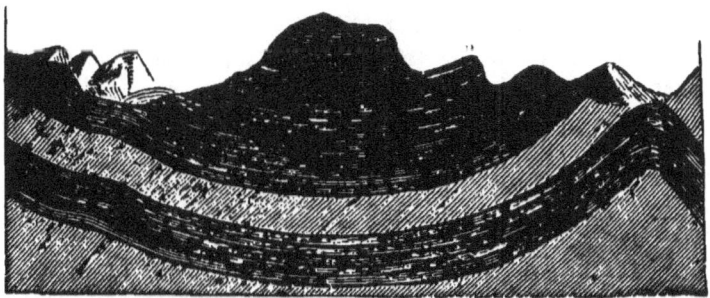

Illustration of "Synclinal" strata.

to strew or spread out) means a single bed of rock, sand, gravel, or clay. When rocks, therefore, lie in parallel layers, they are said to be *stratified*; and when there is no appearance of such arrangement, they are termed *unstratified* STRATA is the plural of STRATUM. A geological FORMATION

U

means an assemblage of rocks which have some character in common, either of age, origin, or composition. Usually such rocks are grouped together into a system by having a great number of fossils common to them. Both in the use of this word, and SYSTEM, however, there is a great deal of looseness. The latter signifies groups of strata which have intimate relations to each other, generally in the order or sequence of their deposition. EPOCH is a word frequently used to express a particular division or period of geological time. It is, therefore, employed as being almost synonymous with " Age " or " Era."

By the term CONFORMABLE, is meant that the strata lie in

Fig. 175.

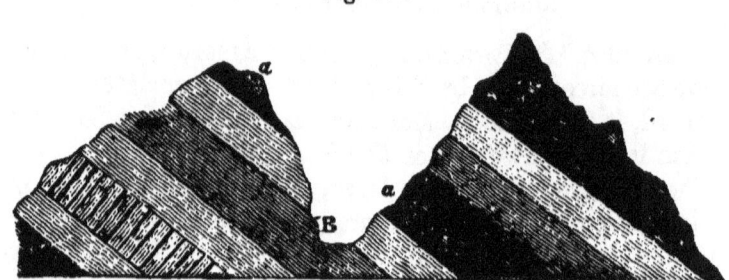

Illustration of " Dip " of strata.

parallel order, one above another. When this is not the case, but when one set of strata, for instance, lies on the upturned edges of the lower, the two are said to be UNCONFORMABLE. Unconformability is generally regarded as a proof of a break in the order of deposition, and therefore of a period of time, sometimes very great, having elapsed in the interval of the formation of the two groups of rocks. By the DIP of a rock is meant the angle or inclination at which the strata slope. This is always measured from the level of the horizon. The opposite of DIP is RISE—a word much in use among coal-miners. They both mean the same thing, only DIP has reference to the position of the observer standing at the surface, who sees the rock sloping away from him—whereas when standing, say at the bottom of a coal-pit, and looking at the same bed, he sees it *rising*. Another frequent word, and one that often causes much trouble to the young geologist, is STRIKE. It means the direction, or line of outcrop, of any stratum. The STRIKE is said to be always at right angles to the *dip*. The best way of illustrating the difference is by supposing the reader to be on the roofs of a row of cottages. Then the ridge running in their length would represent STRIKE, or extension, whilst the sloping of the tiles on either hand would illustrate the *dip*. We have already used the term OUTCROP, and the reader will have little difficulty in understanding what it means. It signifies that part of any inclined stratum of rock which comes to the surface of the ground. Sometimes, if the rock be hard, it thus forms more or less steep cliffs. Hence, whenever a rock thus comes to daylight, it is said to "crop out."

With these few words of explanation for the benefit of the young and unsophisticated student, we bid him "God-speed" to a study which cannot fail to give him health and strength, both of body and mind; or lead him to a wider and broader knowledge of our old world, and of the POWER whose wisdom and love have nursed it from the earliest times!

INDEX.

Acidaspis, 51.
Acrodus, 138.
Actinocrinus, 93.
Æchmodus, 138.
Agates, 17.
Age of Reptiles, 132.
Aleutian islands, 209.
Alum Shales, 128.
Alumina, 5.
American types, 212.
American animals, 219.
Amethyst, 17.
Ammonites, 141, 142, 143, 144, 145, 157, 183, 187.
Amphibia, 164.
Amygdaloid, 1.
Ananchytes, 183.
Annularia, 111.
Anodon, 75, 76.
Anodonta, 75.
Anonas, 198.
Anoplotherium, 200.
Anthracotherium, 201.
Antrim beds, 214.
Appearance of man, 252.
Araucarian pines, 130, 161.
Archægosaurus, 86, 87.
Archæology, 271.
Archæopteryx, 164, 165.
Arctic flora, 249, 252.

Arctic Ice-cap, 215, 221.
Arctic mollusca, 244.
Arencolites, 31.
Argillaceous rocks, 25.
Astartes, 225.
Asterolepis, 67.
Atlantic mud, 179.
Atrypa, 54, 55.
Ancyloceras, 187.
Avicula, 145.
Aymestry limestones, 48.

Baculites, 188.
Bala limestones, 46.
Banksia, 213.
Basaltic columns, 214.
Basaltic rocks, 46.
Basaltic dykes, 214.
Bath Oolite, 150.
Batrachians, 124.
Beaked saurian, 124.
Belemnites, 141, 145, 158, 184, 185.
Bellerophon, 35.
Birds' nests, 162.
Black-lead, 20.
Bovey Tracey lignite, 213.
Boulder, 238.
Boulder clay, 244, 255.
Box-stones, 236.
Brachiopods, 33, 71, 91, 144.

Bracklesham sands, 195.
Breccias, 249.
Brixham cavern, 265.
Brontes, 72.
Bronze period, 269.
Brown coal, 207.
Bryozoa, 225.
Buccinum, 233.
Bulimina, 170.
Bunter sandstone, 116.

Cairngorm, 17.
Calamites, 81, 83, 84, 104, 105, 107.
Calceola, 60.
Cambrian rocks, 24.
Camels, 219.
Camphor-trees, 212.
Caradoc sandstones, 46.
Carboniferous, 80.
Carboniferous limestone, 90, 96.
Carboniferous forests, 106.
Carcharodon, 202.
Cardita, 224.
Cassidaria, 225.
Casts, 35.
Cave Tiger, 235.
Cephalaspis, 68.
Cephalopoda, 35.
Ceratites, 118.
Chæropotamus, 201.
Chain-coral, 54.
Chalcedony, 17.
Chalk, 167, 181, 190.
Chalk foraminifera, 167.
Cheirotherium, 124.
Cheshire meres, 122.
Chipped flints, 263.
Cidaris, 152, 184.
Clay, 193.
Claystone porphyry, 5.

Cleavage, 25.
Clinkstone porphyry, 5.
Club-mosses, 57.
Clymenia, 62, 72.
Coal, 79.
Coccoliths, 179.
Cockles, 185.
Comatula, 138.
Compsognathus, 163.
Conclusion, 291.
Cones, 202.
Coniferous trees, 112.
Contorted rocks, 27.
Coprolites, 229.
Coralline crag, 223, 224, 226, 233.
Corallines, 49.
Coral Rag, 150.
Coral-reefs, 53.
Cornbrash, 150.
Cosmical hypothesis, 2.
Crags, 222.
Cretaceous formation, 80.
Cretaceous sea, 169, 179.
Crinoids, 92.
Cristellaria, 176, 178.
Crystals, 17.
Ctenoid fish, 185.
Cuttle-fish, 35.
Cyathophyllum, 73.
Cycads, 161, 162.
Cycloid fish, 185.
Cyclopteris, 74.
Cyprea Europea, 228.
Cyprina, 224.
Cyrtoceras Murchisoni, 48.
Cystideans, 33, 53.

Dapedius, 138.
Dartmoor granite, 213.
Dawn-animalcule, 18.

Dead Sea, 120.
Dead shells, 173, 174, 177.
Deer, 218.
Deinosauria, 162.
Deinotheria, 218.
Deltas, 216.
Dendritic crystals, 8, 9.
Dentalina, 167, 168.
Denudation, 12.
Devonian rocks, 62.
Dichobune, 201.
Dicynodonts, 127.
Diluvium, 238.
Dirt-beds, 161.
Downton sandstones, 48.
Dryandroides, 213.
Dryopithecus, 218.
Dudley strata, 45.

Earthquakes, 189.
Echinodermata, 183.
Emarginula, 228.
Encrinites, 117.
England joined to continent, 249.
English chalk, 168, 169.
Elephants, 231, 235, 249.
Eocene, 80, 194, 195, 202, 209.
Eozoon, 18.
Etna, 237.
Eucalyptocrinus, 93.
Euomphalus, 53.
Eurypterus, 66.
Evergreen plants, 212.
Excess of salt, 121.
Extracrinus, 138, 141.

Fairy loaves, 183.
Fan palms, 213.
Fascicularia, 225.
Faults, 113.

Favosites, 54, 72, 73.
Feather-star, 138.
Felspar, 3.
Felspar porphyry, 5.
Fig-trees, 213.
Flabellarias, 213.
Flabellina, 177, 178.
Flint, 17, 180, 181.
Flint arrow-head, 264.
Flint bands, 180.
Flint cores, 266.
Flint implements, 260, 261, 265, 267.
Flint nodules, 180.
Flying lizards, 161.
Foldings of rocks, 26.
Foot-print impressions, 119.
Foot-prints of birds, 125.
Foraminifera, 167, 168, 169, 170, 171, 172, 173, 174, 175, 176, 177, 178.
Forest bed, 234, 235, 241.
Forest marble, 150.
Formation of lakes, 251.
Fossil animalculæ, 175.
Fossil butterflies, 216.
Fossil caddis worms, 216.
Fossil corals, 71, 225.
Fossil eggs, 163.
Fossil feathers, 165.
Fossil ferns, 106, 131.
Fossil ferns, 211.
Fossil flora, 211.
Fossil flowers, 211.
Fossil fruits, 197.
Fossil insects, 216.
Fossil Oolitic wood, 160.
Fossil pitch, 129.
Fossil tree fern, 109.
Fossil turtles, 203.

INDEX.

Four-horned deer, 219.
France, Central, strata in, 216.
Free-stone, 61.
Fresh-water lakes, 215.
Frondicularia, 168, 177.
Fucoids, 61.

Ganoid fishes, 66, 198.
Gasteropoda, 35.
Gaudryina, 169.
Gault, 187.
German Ocean, 233, 241.
Giraffe, 219.
Glacial beds, 232, 255.
Glacial boulders, 239.
Glacial period, 194, 239, 259.
Glaciers, 242, 244, 245.
Globigerina, 173, 174, 176.
Glutton, 249.
Gneiss, 18.
Gold, 57.
Goniatites, 90.
Granite, 4.
Granite of Cornwall, 10.
Granite of Scotland, 10.
Granite nucleus, 12.
Granite bosses, 27.
Graptolites, 40, 49, 50.
Gravel pit, 254, 272.
Gravels, 259.
Greek fable, 272.
Greenland lignite beds, 214.
Greenland ice, 243.
Growth of civilisation, 271.
Gryphea, 144, 153.
Gulf-stream, 245.
Gypsum, 119.
Gyroceras, 90.

Hafted stone celt, 270.

Hairy rhinoceros, 249.
Hamites, 188.
Halysites, 54.
Harlech grits, 30.
Headon series, 195.
Hearts, 183.
Heliolites, 72.
Hipparion, 219.
Hippodium, 145.
Hippopotamus, 218.
Homanolotus, 72.
Holoptychius, 68.
Hornblende, 3.
Horse-tails, 103, 104.
Hybodus, 138.
Hyænodon, 201.
Hylæosaurus, 187.
Hymenocaris, 33.

Ice action, 249.
Icebergs, 243.
Iceland lignite beds, 214.
Iceland, fossil plants in, 214.
Ice-sheet, 250.
Ichthyosaurus, 132, 133. 134.
Ichthyocrinus, 93.
Ideal landscape of Triassic period, 114.
Ideal landscape of Oolitic epoch, 147.
Ideal landscape of Oolitic epoch, 206.
Ideal landscape of Quaternary period, 254.
Iguanodon, 187.
Illænus, 52.
Inferior Oolite, 150.
Influence of earth's heat, 23.
Irish elk, 250, 251.
Iron age, 269.

INDEX.

Iron-stone, 156.

Japanese Islands, 209.
Jasper, 17.
Jet, 128, 149.
Joints, 26.
Jordan river, 121.
Jura Alps, 165.

KAOLIN, 4.
Kelloway rock, 150.
Kent's cavern, 265.
Keuper beds, 116, 119.
Keuper sea, 121.
Kilkenny sandstones, 74.
Kimmeridge clay, 150.
King-crab, 49.

Labyrinthodonts, 86, 124, 125.
Lake Dwellings, 269.
Lamp-shells, 144.
Land ice, 242.
Landscape of Carboniferous period, 78.
Last active volcanoes, 214.
Laurentian, 17.
Lava, 9.
Leiodon, 185.
Lemming, 249.
Lepidodendron, 75, 82, 84, 85, 99, 103.
Lepidosteus osseus, 66.
Lepidostrobus, 82, 99.
Lepidotus, 138.
Leptæna, 55.
Liassic formation, 80, 128.
Lignite beds, 201, 207, 211.
Lily encrinite, 117.
Limpet shells, 229.
Limestone, 37.

Lingula, 33.
Lingula flags, 30, 33.
Lingula Lewisii, 45, 55.
Lituola, 177.
Llandeilo flags, 46.
London clay, 194, 229.
Long-armed monkey, 220.
Longmynd rocks, 31.
Lower Cambrian, 31, 32.
Lower Cretaceous, 188.
Lower green sand, 187.
Lower Oolite, 150.
Lower Silurian, 47.
Ludlow beds, 48.
Ludlow bone-bed, 57.
Lump of clay, 193, 196, 197, 198.

Machairodus, 218.
Mactra, 230.
Magnesian limestone, 115.
Magnolias, 213.
Mammoth, 248, 263.
Man's appearance, 252, 259.
Mastodon, 217, 218.
Marginulina, 168.
Marsupials, 126, 131.
May Hill sandstones, 47. 52.
Mechanical origin of rocks, 25, 38.
Megalichthys, 97.
Megalodon, 72.
Megalosaurus, 163, 187.
Metal lodes, 57.
Metamorphic rocks, 18.
Meteorological agencies, 38.
Mesozoic epoch, 117.
Mica, 3.
Mica schist, 10.
Micraster, 183.
Microlestes, 126, 127.
Microscopic section, 180, 181.

Microscopic structure of fossil wood, 112.
—— of coal, 112.
Middle drift, 255.
Middle Silurian, 47, 52.
Middle Oolite, 150.
Minute fossils, 177.
Miocene, 80, 206, 208, 210.
Miocene forests, 210.
Miocene plants, 211.
Missing links, 163, 165, 201, 219.
Mitres, 202.
Moel Tryfaen, 245.
Molasse, 209.
Monkeys, fossil, 217.
Mososaurus, 185, 186.
Mountain-chains, 11.
Murchisonia, 52.
Muschelkalk, 116, 118.
Muskdeer, 249.
Mussels, 236.
Mya, 231, 232.
Mylodon, 222.

Natica, 228.
Nautilus, 53, 141, 143, 202.
Neocomian beds, 187.
Neolithic implements, 264, 267.
Neptunian controversy, 2.
Neuropteris, 106.
Newer Stone age, 266.
New Red Sandstone, 116.
Nipadites, 197.
Nodosaria, 167.
Norfolk cliffs, 244.
Norfolk forest-bed. 234.
North Atlantic, 145.
Northern drift, 239.
Northern shells, 231.
Norway spruce pine, 234, 242.

Norwich Crag, 223, 230, 234, 241.
Nummulites, 195, 196.

Obolus, 55.
Older Pliocene, 223.
Oldhamia, 32.
Old Red Sandstone, 62, 70.
Onchus, 69.
Oolitic animals, 166.
Oolitic period, 80, 149.
Oolitic strata, 160.
Ophioderpeton, 86.
Opossum, 218.
Orthis, 55.
Orthoceratites, 35, 49, 90.
Osmunda, 75.
Osteolepis, 68, 69.
Oxford clay, 150.
Oysters, 236.

Palæolithic flint implements, 266.
Palæolithic period, 263.
Palæoniscus, 98.
Palæopteris, 74.
Palæopyge, 32.
Palæotheria, 198, 199.
Palæozoic brachiopods, 55.
Palæozoic Encrinites, 93.
Paludina, 153, 155.
Paradoxides, 34.
Paramoudra, 182.
Parkia, 65.
Pebbles, 256.
Pecopteris, 109.
Pecten, 53.
Pectunculus, 224, 228.
Pegmatite, 5.
Pentacrinus, 139.
Pentamerus, 53, 54, 55.
Pentremites, 95.

Percolation, 35.
Permian breccias, 115.
Permian formation, 115.
Petroleum, 41.
Phillipsia, 92.
Phlebopteris, 156.
Phosphates, 21.
Phosphate of lime, 188.
Phosphatic nodules, 229.
Phragmacones, 184.
Piece of chalk, 167.
Pilton group, 71.
Pitchstone, 8.
Placoid fishes, 69.
Plagiostoma, 145.
Planorbis, 153, 230.
Platycrinus, 83.
Pleistocene, 231.
Plesiosaurus, 132, 135.
Pliocene, 80, 215, 223, 226, 236, 237.
Pliopithecus, 218.
Plumbago, 20.
Plutonic controversy, 2.
Polished Celts, 269.
Polypothecia, 180, 181.
Porphyrine (artificial), 9.
Porphyry, 5.
Portland Stone, 150.
Pot-stones, 182.
Primary rocks, 15.
Primitive rocks, 6.
Producta, 54, 115.
Proteacea, 213.
Protogine, 5.
Protozoic rocks, 28.
Psilophyton, 75.
Psychometry, 59.
Pterichthys, 67, 68.
Pterodactyles, 132, 136, 137, 153.

Pterophyllum, 131.
Pterygotus, 65.
Purbeck marble, 146, 153, 159.
Purbeck series, 150.
Pyrula, 225.
Pyrulina, 168.

Quartz, 3, 14.
Quartz crystals, 7.
Quartzite, 16.
Quaternary, 80.

Rain-drop pittings, 119.
Red Crag, 223, 229, 230, 241.
Red Crag, shells of, 229.
Reindeer, 249, 266.
Reindeer period, 266.
Reptile-like birds, 164.
Restored Belemnite, 185.
Restored Calamites, 107.
Restored Lepidodendron, 99.
Rhætic formation, 126.
Rhinoceros, 234, 236, 249.
Rhodocrinus, 93.
Rhynchonella, 54, 145, 184.
Rhynchosaurus, 124.
Ripple-marks, 31, 32, 89, 119, 154.
River-gravels, 255.
Rostellaria, 202.
Rock-oil, 41.
Rock-salt, 115.
Rosalina, 172.
Rotalina, 173, 175, 177.

Sagenaria, 75.
Sagrina, 174.
Sahara sea, 251.
Salisburia, 215.
Sandstone, 58.
Scales of Carboniferous fish, 98.

Scaphites, 188.
Sea-urchins, 183.
Section of fossil wood, 103.
Segregation of metals, 57.
—— of minerals, 57.
Semnopithecus, 218.
Sequoias, 213.
Sewalik hills, deposit of, 219.
Sharks, 185, 203.
Sicilian beds, 236.
Sigillaria, 84, 85, 102, 103.
Silica, 3, 17, 179, 181.
Silurian rocks, 39.
—— limestones, 41, 45.
Single corals, 95.
Siphuncle, 143.
Sivatherium, 219.
Slate rocks, 24.
Smilax, 212.
Spiculæ, 182.
Spirifera, 54, 55, 72, 91, 144.
Spirorbis, 84.
Spondylus, 185.
Stalactite, 265.
Stalagmite, 265.
Star-fish, 53.
Stauria, 71.
Stone Celt, 270.
Stonesfield slate, 150, 163.
Stringocephalus, 72.
Stromatopora, 72.
Strophomena, 54, 55.
Structure of Pentremites, 96.
Sub-tropical climate, 190.
Suffolk strata, 223.
Sun-cracks, 32, 89, 119, 120.
Sun-corals, 72.

Tapiroid animals, 199.
Tapirs, 218.

Tarannon shales, 47.
Taxocrinus, 93.
Teeth of Carboniferous fish, 98.
Tellina Balthica, 244.
Tellina crassa, 228.
Temnechinus, 224.
Temperature of Miocene Period, 209.
Terebella, 202.
Terebratula, 54, 145, 151, 184.
Terraces, 255.
Terraced hills, 256, 257.
Tertiary epoch, 191, 194.
Textularia, 169.
Theca, 34.
Thickness of lias, 130.
Thunder-bolts, 144, 145, 158, 184.
Topaz, 17.
Tortoise, huge, 219.
Transition rocks, 18.
Transverse section of calamite, 83.
Transverse section of fruit of calamite, 105.
Tree-apes, 218.
Tree-ferns, 105.
Trias, 116.
Triassic Period, 114.
Trigonia, 154.
Trilobites, 34, 40, 41, 49, 51, 72.
Trogontherium, 235.
Truncatulina, 171.
Turrilites, 188.
Turritella, 195.
Turtles, 203.
Typhis, 202.

Unconformable rocks, 22, 119.
Upper Crag, 233.
Upper Eocene, 204.
Upper green sand, 188.

Upper Oolite, 150.
Upper Silurian, 48, 53, 54.

Valley-gravels, 255, 263.
Vancouver's Island, 209.
Verneuilina, 170.
Vertebrata, 54.
Vertical section of calamite, 81.
Vertical section of fruit of calamite, 104.
Volutes, 202, 226.

Wart-hogs, 218.

Wealden strata, 186, 187.
Wenlock group, 48.
Wenlock limestones, 42, 48.
Wentle-traps, 230.
White whelk, 233.
Woolhope beds, 48.
Woolly-haired elephant, 263.
Woolly-haired rhinoceros, 263.
Worm-tracks, 32.

Zamias, 161.
Zaphrentis, 72.

THE END.

LONDON:
PRINTED BY WILLIAM CLOWES AND SONS,
STAMFORD STREET AND CHARING CROSS.

Interesting and Useful Works

PUBLISHED OR SOLD BY

OBERT HARDWICKE,

192, PICCADILLY,

W.

Y WORKS IN THIS CATALOGUE MAY BE HAD FROM ANY BOOKSELLER.

INDEX.

Name	Page
Huxley, Prof., F.R.S.	39
Jones, Prof. T. R., F.G.S.	43
Jones, Dr. W. H., M.R.C.S.	35
K. K. (illuminated books)	30
Kinahan, G. H.	17
Langley, J. Baxter, M.R.C.S., F.L.S.	35
Lankester, Dr., F.R.S.	13, 15, 29, 30
Lankester, E. Ray	42
Lankester, Mrs.	5, 11, 12
Lawson, Dr.	18, 29, 35, 41
Le Hardy, Miss	27
Leighton, Rev. W. A., M.A.	40
Lewes, G. H.	42
Liebig, Baron	36
Lord, J. K.	16
Lowth, G. T.	30
Lukis, Captain J. H.	27
Mackie, S. J., F.G.S.	43
Marsh, Dr. J. C. L., M.R.C.P.	27
Masters, Dr. M., F.L.S.	40
Meyen, Prof.	40
Miller, W. A., M.D., V.P.R.S.	46
Milne, Alex., M.D.	36
Milton, J. L.	18, 36
Mivart, Geo. F.G.S.	46
Monckhoven, D. Van, Ph.D.	19
Morris, Dr. E., F.R.C.S.	36
Morris, Prof., F.R.S.	47
Nelson, David, M.D.	17
Norton, A. T., F.R.C.S.	36
Nunn, T. W., F.R.C.S.	37
Ogle, W., M.D.	46
Parker, W. K., F. S.	40
Patterson, Robt., F.R.S.	43
Pengelly, W., F.R.S.	47
Proctor, R. A., F.R.A.S.	20, 44
Procter, Wm., M.D.	37
Pullar, Alfred, M.D.	36
Ransonnette, Baron	28
Robson, J. E.	15
Richardson, Dr. B. W., F.R.S.	43
Sanderson, Dr. J. B., F.R.S.	37
Samuelson, James	42
Sawyer, James M. B.	37
Shaible, Dr. C. H., Ph.D.	28
Schleiden, Prof.	12
Sclater, Dr., F.R.S.	40
Scott, John	39
Seemann, Dr. B., Ph.D., F.L.S., F.R.G.S.	43
Sheen, J. R.	27
Sheppard, Dr. E.	27
Simmonds, P. L.	21
Sims, Dr. J. Marion	37
Smith, David, M.D.	37
Smith, John, A.L.S.	11
Smith, W. G., F.L.S.	13
Sorby, H. C., F.R.S.	44
Sowerby, J. E.	5, 12
Spear, J.	24
Spicer, Rev. W. W., M.A.	14
Steenstrup, Prof.	39
Strickland, H. E., F.R.S.	...
Syme, J. T. B., F.L.S.	5
Symonds, Rev. W. S.	16, 43
Tate, R., F.R.G.S.	13
Taylor, Dr. A. S., F.R.S., F.R.C.P.	31
Taylor, J. E., F.G.S.	17, 19
Teevan, W. F., F.R.C.S.	38
Trimen, H., M.B., F.L.S.	11
Trousseau, Prof. A.	38
Turner, Mansfield	24
Unwin, W. C.	46
Vincent, John	19
Voelcker, Prof., F.C.S.	47
Waite, S. C.	21
Walford, Rev. E.	25
Walker, C. V., F.R.S.	43
Whitley, N., C.E.	46
Williamson, Prof., F.R.S.	39
Winn, J. M., M.D.	38
Wood, Jno., F.R.C.S.	37
Zander, Dr. A.	33
Zerffi, Prof.	30

A CATALOGUE OF BOOKS

PUBLISHED OR SOLD BY

ROBERT HARDWICKE,

192, PICCADILLY, LONDON, W.

Sowerby's English Botany:

Containing a Description and Life-size Drawing of every British Plant. Edited and brought up to the present standard of scientific knowledge, by T. BOSWELL SYME, LL.D., F.L.S., &c. With Popular Descriptions of the Uses, History, and Traditions of each Plant, by Mrs. LANKESTER, Author of "Wild Flowers Worth Notice," "The British Ferns," &c. The Figures by J. E. SOWERBY, JAMES SOWERBY, F.L.S., J. DE C. SOWERBY, F.L.S., and J. W. SALTER, A.L.S.

"Under the editorship of T. Boswell Syme, F.L.S., assisted by Mrs. Lankester, 'Sowerby's English Botany,' when finished, will be exhaustive of the subject, and worthy of the branch of science it illustrates...... In turning over the charmingly executed hand-coloured plates of British plants which encumber these volumes with riches, the reader cannot help being struck with the beauty of many of the humblest flowering weeds we tread on with careless step. We cannot dwell upon many of the individuals grouped in the splendid bouquet of flowers presented in these pages, and it will be sufficient to state that the work is pledged to contain a figure of every wild flower indigenous to these isles."—*The Times.*

"Will be the most complete Flora of Great Britain ever brought out. This great work will find a place wherever botanical science is cultivated, and the study of our native plants, with all their fascinating associations, held dear."—*Athenæum.*

"Nothing can exceed the beauty and accuracy of the coloured figures. They are drawn life-size—an advantage which every young amateur will recognize who has vainly puzzled over drawings in which a celandine is as big as a poppy—they are enriched with delicate delineations of fruit, petal, anther, and any organ which happens to be remarkable in its form—and not a few plates are altogether new...... A clear, bold, distinctive type enables the reader to take in at a glance the arrangement and divisions of every page. And Mrs. Lankester has added to the technical description by the editor an extremely interesting popular sketch, which follows in smaller type. The English, French, and German popular names are given, and, wherever that delicate and difficult step is at all practicable, their derivation also. Medical properties, superstitions, and fancies, and poetic tributes and illusions follow. In short, there is nothing more left to be desired."—*Guardian.*

"Without question, this is the standard work on Botany, and indispensable to every botanist...... The plates are most accurate and beautiful, and the entire work cannot be too strongly recommended to all who are interested in botany."—*Illustrated News.*

*** Subscribers to this great National Undertaking may commence at any time without buying all the back numbers at once.
Prospectuses and Specimens gratis.

Sowerby's English Botany, Volume 1 contains

Aconite	Coralwort	Ladies' Smocks	Pheasant's eyes
Alyssums	Corydalises	Larkspurs	Poppies
Anemones	Cresses	Lilies	Radishes
Awlwort	Crowfoots	Madwort	Rape
Banebury	Cuckoo flower	Marigold	Rockets
Barberry	Earthsmoke	Mayflower	Rues
Barrenwort	Figwort	Monkshood	Scurvy grasses
Bearsfoot	Flixweed	Mousetails	Spearworts
Buttercup	Fumitories	Mustards	Stocks
Cabbages	Gilliflowers	Nailwort	Toothwort
Caltrops	Headwark	Nasturtiums	Traveller's joy
Candytuft	Hellebores	Navews	Turnips
Celandines	Horseradish	Old man's beard	Wallflowers
Charlocks	Hutchinsia	Pasque flower	Whitlow grasses
Coleseed	Jack-by-the-hedge	Peony	Woad
Colewort	Kingcup	Pepperworts	Wolfsbane
Columbine			

All the Plants ranked under the orders Ranunculaceæ, Berberidaceæ Nymphæaceæ, Papaveraceæ, and Cruciferæ.

Sowerby's English Botany, Volume 2 contains the

Aaron's beard	Cyphel	Pansies	Spindle tree
Alder	Dyer's weed	Pearlworts	Spurreys
All-seed	Flaxes	Pinks	Stitchworts
Balsams	Fuller's herb	Purslane	Storks' bills
Bladder-nut	Heartsease	Ragged Robin	Sun dews
Bruisewort	Herb Roberts	Red rot	Sun rose
Buckthorns	Holly	Rocket	Sycamore
Campions	Limes	Rock roses	Tamarisk
Carnation	Limewort	Rose of heaven	Tutsan
Catchflys	Lychnis	St. John's worts	Violets
Chickweeds	Mallows	Sandworts	Water blinks
Claytonia	Maples	Sea heath	Waterworts
Corn cockle	Mignonettes	Soapwort	Wild Williams
Crane's bills	Milkworts	Sorrels	Yarr
Cuckoo flower	Moenchia		

All the Plants ranked under the orders Resedaceæ, Cistaceæ, Vi laceæ, Droseraceæ, Polygalaceæ, Frankeniaceæ, Carophyllaceæ, P tulacaceæ, Tamariceæ, Elatinaceæ, Hypericaceæ, Malvaceæ, Tiliace Linaceæ, Geraniaceæ, Ilicineæ, Celastraceæ, Rhamnaceæ, Sapindace

Sowerby's English Botany, Volume 3 contains the

Agrimonies	Burnets	Gean	May
Apples	Cherries	Gorse	Meadow-sweet
Avens	Cinquefoils	Greenweeds	Medicks
Beams	Cloudberry	Hawthorns	Medlar
Birdsfoots	Clovers	Honeystalks	Meliot
Blackberry	Cotoneaster	Ladies-finger	Mountain-ashes
Blackthorn	Dewberry	Lady's mantles	Nonsuchs
Brambles	Dropwort	Lamb-toe	Oxytropis
Briars	Eglantine	Liquorice	Parsley piert
Broom	Fenugreek	Lucernes	Pear
Bullace	Furzes	Marl-grass	Peas

Robert Hardwicke, 192, *Piccadilly.*

owerby's English Botany, Volume 3 (*continued*) contains the

um	Saintfoin	Sweetbriars	Vetches
een of the mea-	Service-trees	Tares	Vetchlings
dow	Shamrock	Tongue under	Waxen wood
spberries	Sibbaldia	tongue	Whins
st-harrows	Silver-weed	Tormentils	Whitebeams
ebuck-berry	Spiræa	Trefoils	Whitethorns
ses	Strawberries		

All the Plants ranked under the orders Leguminiferæ and Rosaceæ.

werby's English Botany, Volume 4 contains the

exanders	Earth-nuts	Isnardia	Pennywort
gelicas	Elders	Ivy	Purslane
ise	Eryngo	Lamb's-lettuces	Rose-root
trantia	Evening primroses	Linnæa	Samphires
ld-money	Fennels	Livelong	Sanicle
y, rose	Golden moss	London Prides	Saxifrages
dstraws	Gooseberry	Loose strife	Scabiouses
adder-seed	Goosegrass	Lovage	Squinancywort
yony	Goutweed	Madders	Stonecrops
raways	Grass of Parnassus	Mare's tail	Stoneworts
rrots	Grass-poly	Masterwort	Sulphur-wort
lery	Guelder-rose	Meal tree	Teasels
eese-rennet	Hare's-ears	Milfoils	Thorough-wax
ervils	Hartwort	Mistletoe	Tillæa
cely	Hemlocks	Moschatel	Valerians
riander	Herb, Bennet	Mugwort	Venus's comb
rnel	Herb, Gerard	Navelwort	Wayfaring tree
rrants	Hog's fennels	Nightshades	Willow herbs
newort	Holley, sea	Orpines	Woodbines
gwood	Honeysuckle	Parsleys	Woodruffs
opworts	House-leek	Parsnips	

All the Plants ranked under the orders Lythraceæ, Onagraceæ, curbitaceæ, Grossulariaceæ, Crassulaceæ, Saxifragaceæ, Umbelliferæ, raliaceæ, Cornaceæ, Loranthaceæ, Caprifoliaceæ, Rubiaceæ, Valerianeæ, and Dipsaceæ.

werby's English Botany, Volume 5 contains the

ter	Elecampane	Heliotrope	Salsify
e-bottle	Everlastings	Hemp-agrimony	Samphire
rdocks	Feverfew	Inula	Saw-worts
r-Marygolds	Fleabanes	Knapweeds	Sneeze-worts
tter-bur	Fleaworts	Lettuces	Southernwood
t's Ears	Galinsoga	Leopards'-banes	Sow-thistle
amomiles	Goats-beards	Marigold	Spikenard
ltsfoots	Golden-rods	Mayweeds	Star-thistles
rn-flower	Goldylocks	Mugwort	Succorys
ton-weed	Groundsels	Nipple-wort	Tansy
dweeds	Hawk-bits	Ox-eye	Thistles
sy	Hawks'-beards	Ox-tongue	Wormwoods
delions	Hawkweeds	Ragworts	Yarrows

All the Plants ranked under the order Compositeæ.

Sowerby's English Botany, Volume 6 contains the

Andromeda	Bunny	Henne-belle	Rosemary
Arbutus	Butter-and-eggs	Higtaper	Sheep's-bit
Ashes	Canterbury-bell	Jacob's ladder	Snapdragons
Azalea	Centaurys	King's-taper	Speedwells
Bartsias	Cicendias	Ling	Strangle-weed
Bearberrys	Convolvulus	Lobelias	Strawberry-tree
Belladonna	Cow wheats	Louseworts	Tea-plant
Bell-flowers	Cranberry	Menziesias	Thorn-apple
Betonys	Devale	Moneywort	Throat-wort
Bilberrys	Dodders	Monkey-flower	Toadflaxes
Bindweeds	Euphraisie	Mother-of-thousands	Tooth-wort
Bird's-nest	Eyebrights		Torch-blade
Bitter-sweet	Felwort	Mudwort	Valerian
Bleaberry	Figworts	Mulleins	Venus's looking glass
Bog bean	Fleullins	Nightshades	
Broom-rapes	Foxglove	Pennyweed	Vervian
Brooklime	Gentians	Periwinkles	Violet
Brown-rapes	Hag-taper	Primprint	Whortleberry
Brownworts	Hare-bell	Privets	Witches' thimbl
Buck beans	Heaths	Rabbit's-mouth	Winter-greens
Bull-dogs	Heather	Ramplons	Yellow-rattles
Bullock's-wort	Henbane		

All the Plants ranked under the orders Campanulaceæ, Ericaceæ, Jasminaceæ, Apocynaceæ, Gentianaceæ, Polemoniaceæ, Convolvuceæ, Solanaceæ, Scrophulariaceæ, Orabanchaceæ, and Verbenaceæ.

Sowerby's English Botany, Volume 7 contains the

Alkanets	Comfreys	Knawels	Pimpernels
Amaranth	Cowslips	Lavenders	Plantains
Archangel	Creeping Jenny	Loosestrifes	Primroses
Balms	Cyclamen	Lungworts	Rib-grass
Basil	Dead-nettles	Madwort	Ruptureworts
Betony	Forget-me-nots	Marjoram	Saltwort
Bladderworts	Germanders	Mints	Skull-caps
Borage	Gipseyworts	Moneywort	Sea lavenders
Brook-weed	Gromwells	Motherwort	Self-heal
Bugles	Ground ivy	Nettles	Strapwort
Buglosses	——— pine	Oxlips	Thrifts
Butterworts	Hemp-nettles	Oyster-plant	Thymes
Calamints	Horehounds	Pennyroyal	Water violets
Chickweed	Hounds' tongues	Peppermint	Woundwort
Clarys	Illecebrum		

All the Plants ranked under the orders Labiatæ, Boraginaceæ, Lentibulariaceæ, Primulaceæ, Plumbaginaceæ, Plantaginaceæ, Paronchiaceæ, and Amarantaceæ.

Sowerby's English Botany, Volume 8 contains the

Alder	Bistorts	Elms	Junipers
Allgocd	Box	Fir	Knotgrasses
Asarabaca	Buckthorn	Goosefoots	Mercury
Aspen	Buckwheats	Hazel	Mezereon
Beech	Chestnut	Hemp	Myrtle
Beet	Crowberry	Hop	Nettles
Birches	Docks	Hornbeam	Oaks
Birthworts	Dog's mercurys	Hornworts	Oraches

Sowerby's English Botany, Volume 8 *(continued)* contains the

Osiers	Poplars	Samphires	Spurges
Pellitory of the wall	Purslane	Seablites	Starworts
Pepper	Rhubarb	Sea beet	Willows
Persicarias	Sallows	Sorrels	Yew
Pine	Saltworts		

All the Plants ranked under the orders Chenopodiaceæ, Polygonaceæ, Eleganaceæ, Thymelaceæ, Santalaceæ, Aristolochiaceæ, Empetraceæ, Euphorbiaceæ, Callitrichaceæ, Ceratophyllaceæ, Urticaceæ, Amentiferæ, and Coniferæ.

Sowerby's English Botany, Volume 9 contains the

Arrow-grasses	Epipogium	Lily of the valley	Snowdrop
Arrow-head	Flowering rush	Lloydia	Snowflakes
Asparagus	Fritillary	Martagon lilies	Solomon's seals
Asphodels	Frog-bit	Naias	Squils
Bryony	Garlics	Narcissuses	Star of Bethels
Bur-reeds	Gladiolus	Orchises	Sweet-flag
Butchers' broom	Grass-wracks	Pondweeds	Thrumwort
Cats' tails	Helleborine	Ramsons	Trichonema
Chives	Herb-Paris	Ruppias	Tulip
Coral-root	Hyacinths	Saffrons	Tway-blades
Crocuses	Irises	Scheuchzeria	Water plantains
Cuckow-pints	Lady's slipper	Semethis	Water-soldier
Daffodils	Lady's tresses	Sisyrinchium	Water thyme
Duckweeds	Leeks	Smilicina	

All the Plants ranked under the orders Typhaceæ, Araceæ, Lemnaceæ, Naiadaceæ, Alismaceæ, Hydrocharidaceæ, Orchidaceæ, Iridaceæ, Amaryllidaceæ, Dioscoriaceæ, and Liliaceæ.

Sowerby's English Botany, Volume 10 contains the

Pipewort	Sharp-flowered Rush	Blysmuses	Prickly Sedges
Wood Rushes	Shining-fruited Rush	Club Rushes	Grey Sedge
Three-leaved Rush	Lesser-jointed Rush	Bull Rushes	Spiked Sedges
Clustered Rush		Cotton Grasses	Axillary Sedge
Three-flowered Rush		Kobresia	White Sedges
Two-flowered Rush	Capitate Rush	Dioecious Sedges	Haresfoot Sedge
Sea Rushes	Toad Rushes	Flea Sedge	Black Sedge
Common Rush	Mud Rush	Rock Sedge	Alpine Sedges
Soft Rush	Heath Rush	Few-flowered Sedge	Hoary Sedge
Diffuse Rush	Round-fruited Rush	Curved Sedge	Tufted Sedge
Hard Rush	Brown Cyperus	Marsh Sedges	Mountain Sedges
Northern Rush	Galingale	Brown Sedges	Water Sedges
Thread Rush	Black Schœnus	Sea Sedge	Heath Sedges
Hunt-flowered Rush	Fen Sedge	Panicled Sedges	Mud Sedges
	Beak Sedges	Paradoxical Sedge	Fingered Sedges
		Great Sedge	Silvery Sedges
			And other Sedges

And all Plants ranked under the Orders Juncaceæ and Cyperaceæ.

Sowerby's English Botany, Volume 11 contains the

Alpine Grasses	Cut-Grass	Lyme Grass	Quaking Grass
Barley Meadows	Darnel	Marram	Reeds
Bent Grasses	Dog's-tail Grasses	Mat Grass	Ribbon Grass
Bog Hair Grass	Fesque Grasses	Meadow Grasses	Rye Grasses
Bristle Grasses	Finger Grass	Melic Grass	Small Reeds
Brome Grasses	Fox-tail Grasses	Millet Grass	Timothy Grasses
Canary-Grass	Hair Grasses	Moor Grass	Vernal Grass
Cocks'-foot Grass	Hard Grass	Nit Grass	Wall Barley
Cord Grasses	Hares'-tail Grass	Oats	Whorl Grass
Couch Grasses	Heath Grasses	Oat Grass	Yorkshire Fog
Creeping Grasses	Holy Grass	Panic Grass	

All the Plants ranked under the Order Graminaceæ.

Sowerby's English Botany.

The Prices of the Volumes are:—

	Bound cloth £. s. d.	Half morocco £. s. d.	Morocco elegt £. s. d
Vol. 1. (Seven Parts)....	1 18 0	2 2 0	2 8 6
Vol. 2. ditto	1 18 0	2 2 0	2 8 6
Vol. 3. (Eight Parts)....	2 3 0	2 7 0	2 13 6
Vol. 4. (Nine Parts)	2 8 0	2 12 0	2 18 6
Vol. 5. (Eight Parts)....	2 3 0	2 7 0	2 13 6
Vol. 6. (Seven Parts)....	1 18 0	2 2 0	2 8 6
Vol. 7. ditto	1 18 0	2 2 0	2 8 6
Vol. 8. (Ten Parts)	2 13 0	2 17 0	3 3 6
Vol. 9. (Seven Parts)....	1 18 0	2 2 0	2 8 6
Vol. 10. ditto	1 18 0	2 2 0	2 8 6
Vol. 11. (Six Parts)	1 13 0	1 17 0	2 3 6

Or, the Eleven Volumes, £22. 8s. in cloth ; £24. 12s. in half morocco and £28. 3s. 6d. whole morocco.

Every Known Fern.

Synopsis Filicum; including Osmundaceæ, Schizæaceæ, Martiaceæ, and Ophioglossaceæ, accompanied by Figures representi the essential characteristics of each Genus. By the late Sir W. HOOKER, K.H., and JOHN GILBERT BAKER, F.L.S., Assista Curator of the Kew Gardens. Price £1. 2s. 6d. plain ; £1. coloured by hand.

This is the cheapest and most complete book which describes all the Fe throughout the world. So full and clear are its details that anybody who take the trouble to master a few botanical terms can with certainty name any f he may meet with in any part of the world, whether growing wild or under c tivation in garden or hothouse.

"Sir W. Hooker began collecting a Fern Herbarinm in 1811. For this stu this enthusiastic and illustrious botanist had all his life extraordinary advantag The positions he occupied—first as Professor of Botany at Glasgow, and then Director of the Kew Gardens—were made the most of, not merely by amiabil

ssessed the finest herbarium of ferns, and, thanks to Mr John Smith, the best llection of ferns in cultivation in the world. The names of travellers and botasts to whom he acknowledges obligations fill a page and a half of his preface. 'hen Sir W. Hooker died, his son, Dr. Hooker, requested Mr. J. G. Baker to rry it out, possessing as he did full access to the specimens, and being supplied ith Sir William's notes and annotations. The result is the publication of such work as the author designed—a *vade mecum* for the students, cultivators, and llectors of ferns."—*Athenæum.*

'erns, British and Foreign.

Their History, Organography, Classification, Nomenclature, and Culture, with Directions showing which are the best adapted for the Hothouse, Greenhouse, Open Air Fernery, or Wardian Case. With an Index of Genera, Species, and Synonyms. By JOHN SMITH, A.L.S., late Curator of the Royal Gardens, Kew. With 250 Woodcuts. Crown 8vo. cloth, fully illustrated, price 6s.

Mr. Smith is acknowledged to be one of the first authorities on Ferns, having en engaged nearly half a century in arranging them at Kew.

he Fern Collector's Album.

A descriptive Folio for the reception of Natural Specimens; containing on the right-hand page a description of each Fern printed in colours, the opposite page being left blank for the collector to affix the dried specimen; forming, when filled, an elegant and complete collection of this interesting family of plants. Handsomely bound, price One Guinea, size 11¾ in. by 8¼ in. A Large Edition, size 17½ in. by 11 in., without descriptive Letterpress, One Guinea.

he British Ferns.

(A Plain and Easy Account of). Together with their Classification, Arrangement of Genera, Structure, and Functions, directions for out-door and in-door Cultivation, &c. By Mrs. LANKESTER. Fully illustrated, price 4s. coloured by hand; 2s. 6d. plain.

"Not only plain and easy, but elegantly illustrated."—*Athenæum.*

he Flora of Middlesex.

A Topographical and Historical Account of the Plants found in the County; with Sketches of its Physical Geography and Climate, and of the progress of Middlesex Botany during the last three Centuries. By HENRY TRIMEN, M.B. (Lond.), F.L.S., Botanical Department, British Museum, and Lecturer on Botany, St. Mary's Hospital; and WILLIAM T. THISTELTON DYER, B.A., late Junior Student, Christ's Church, Oxford; Professor of Natural History, Royal Agricultural College, Cirencester. With a Map of Botanical Districts. Crown 8vo. cloth, price 12s. 6d.

The object of this book is to give a complete and accurate catalogue of th plants which have at any time been recorded to grow in Middlesex, either a natives or in a more or less completely naturalized state; to indicate the specia localities where they have been found, and to trace the history of their discovery The existence of many of these is attested only by records in scarce or little known books, or by the original specimens preserved in old collections.

"A carefully and conscientiously compiled volume.... The authors have don their best to render it complete and useful to botanists. The biographical an historical notes and the diligent research displayed amongst scarce publication and British Museum manuscripts, render it manifest that the authors have bee engaged in a labour of love, and in a sympathetic record of the scientific diligenc of not a few unremembered but most devoted botanists...."—*Athenæum.*

"So many tastes and interests are catered for in this volume, that our author are sure to get full meed of praise and thanks from all sorts and descriptions c plant-lovers, and we too are happy to aid in the general chorus."—*Gardener Chronicle.*

The London Catalogue of British Plants.

Published under the direction of the London Botanical Exchang Club, adapted for marking Desiderata in Exchanges of Specimens for an Index Catalogue to British Herbaria; for Indicating th Species of Local Districts; and for a Guide to Collectors, b showing the rarity or frequency of the several species. Secon Edition, 8vo. sewed, price 6d.

Wild Flowers worth Notice.

A Selection from the British Flora of some of our Native Plant which are most attractive for their Beauty, Uses, or Association By Mrs. LANKESTER. Illustrated by J. E. SOWERBY. Fcp. 8v cloth, coloured by hand, price 4s.

"We could while away a long summer day talking of the pleasant things su gested by this little book. Although all intelligent persons cannot become bota nists, not to know the wild flowers of our country is to be ignorant both of ou country and ourselves. And this little book will, as a pocket companion durin holiday rambles—the descriptions and plates being both good—destroy this ign rance in reference to at least a hundred plants. After mastering it, the studer will be not a little astounded at his own learning, when he surveys it in the syst matic chapter of contents."—*Athenæum.*

Microscopic Fungi.

(Rust, Smut, Mildew, and Mould). An Introduction to the Stud of Microscopic Fungi. By M. C. COOKE, Author of "The Britis Fungi." Fcap. 8vo. nearly 300 coloured Figures. 2nd Editio with Appendix of New Species, price 6s.

"There is a thoroughness about Mr. Cooke's writings which always makes h communications welcome. He is not content to gather information from cycl pædias, classify and adapt, and then give a new form to the thoughts of other On the contrary, he strikes out a new course of study, and after a laborio course of analysis, produces an entirely original work, one on which nothin of the kind had been before attempted."—*Wesleyan Times.*

The British Fungi.

(A Plain and Easy Account of). With especial reference to the Esculent and other Economic Species. By M. C. COOKE. New and revised Edition, with Coloured Plates of 40 Species. Fcp. price 6s.

"A very readable volume upon the lowest and least generally understood race of plants. For popular purposes the book could not have been better done."— *Athenæum.*

Mushrooms and Toadstools:

How to Distinguish easily the Difference between Edible and Poisonous Fungi; with two large sheets, containing Figures of 29 Edible and 31 Poisonous Species, drawn the natural size, and coloured from living specimens. By WORTHINGTON G. SMITH, F.L.S., &c. In sheets, with book, price 6s.; on canvas in cloth case for pocket, 10s. 6d.; on canvas, with rollers and varnished, for hanging up, 10s. 6d.

"Before the appearance of Mr. Smith's book scarcely more than three or four kinds were freely accepted by prudent cooks. We have now before us the list of esculent mushrooms recommended, nay, warranted by Mr. Smith, of all of which he and his friends have freely partaken, with the most satisfactory results, and many of which are far more delicious in flavour and toothsome in substance than the meadow and horse mushrooms, generally known as the mushrooms of commerce."—*Pall Mall Gazette.*

The British Reptiles.

A Plain and Easy Account of the Lizards, Snakes, Newts, Toads, Frogs, and Tortoises indigenous to Great Britain. By M. C. COOKE, Author of "The British Fungi," &c. Fcap. 8vo. cloth, fully illustrated, 4s. plain; 6s. coloured.

British Mollusks;

Or, Slugs and Snails, Land and Fresh-water. A Plain and Easy Account of the Land and Fresh-water Mollusks of Great Britain, containing Descriptions, Figures, and a Familiar Account of the Habits of each Species. By RALPH TATE, F.R.G.S. Fcap. 8vo. cloth, fully illustrated, 4s. plain; 6s. coloured.

Schleiden's Principles of Scientific Botany.

Or, Botany as an Inductive Science. Translated by Dr. LANKESTER. Hundreds of Woodcuts, and 6 pages of Figures, beautifully engraved on Steel. Demy 8vo. price 10s. 6d.

Every Botanical Library should possess this work, as it contains the principles upon which all structural botany is based.

A Manual of Structural Botany.

By M. C. COOKE. New Edition, with Chemical Notation. Illustrated by more than 200 Woodcuts, price 1s.; bound, 1s. 6d.

"Condensed yet clear, comprehensive but brief."—*Globe.*

"We are confidently able to recommend the little volume to public favour, its very low price bringing it within the range of all purchasers."—*Era.*

A Manual of Botanic Terms.

By M. C. COOKE. New Edition. Greatly enlarged and including the recent Teratological terms. With more than 300 illustrations. Fcp. 8vo. cloth, price 2s. 6d.

"We do not hesitate to say that by a careful use of this book a sound knowledge of the theoretical portion of botany may be obtained without tedious labour."—*Mining Journal.*

The Collector's Handy-Book of Algæ, Diatoms, Desmids, Fungi, Lichens, Mosses, c.

With Instructions for their Preparation and the Formation of an Herbarium. Translated and Edited by the Rev. W. W. SPICER, M.A. With 114 Illustrations. Fcp. 8vo. cloth, price 2s. 6d.

"The publisher of this work well deserves the thanks of all amateur naturalists, and especially of those who have not an unlimited amount of money to expend on their favourite pursuits. The practical hints for collecting, mounting, and preserving, are so varied and abundant that the collector can desire no better *vade mecum.*"—*Naturalists' Circular.*

Economic Products from the Vegetable Kingdom.

Arranged under their respective Natural Orders, with the Names of the Plants, and their parts used in each case. Demy 8vo. price 1s. 6d.

Collection Catalogue for Naturalists.

A Ruled Book for keeping a permanent record of Objects in any branch of Natural History, with Appendix for recording interesting particulars, and lettered pages for general Index. Strongly bound, 200 pages, 7s. 6d.; 300 pages, 10s.; and 2s. 6d. extra for every additional 100 pages. Working Catalogues, 1s. 6d. each.

A really good catalogue, upon a simple and inexpensive plan, has long been a desideratum with naturalists and collectors. Few will originate or prepare a catalogue for themselves, and it is a frequent matter of regret, that large and otherwise valuable collections are rendered comparatively useless for want of such a history. Again, how many naturalists pass through life, enriched with knowledge gleaned from Nature's inexhaustible store, but whose knowledge, alas! dies with them, for they leave behind them no record with their collections. This catalogue commends itself to all naturalists, whether they be ornithologists, entomologists, botanists, conchologists, or mineralogists.

In addition to the ruled and printed form of catalogue, there is a ruled and paged appendix, and an Index lettered and ruled. The appendix is intended to receive such notes and observations as are too long to be recorded in the column for remarks; and the index at the end of the work is intended to furnish an easy

Botanical Labels,

(A Series of.) For labelling Herbaria adapted to the names in the London Catalogue of Plants and the Manuals of Professor Babington and Dr. Hooker, with extra labels for all new species and varieties recorded in the recent volumes of the Journal of Botany, and the Exchange Club Reports. By JOHN E. ROBSON. With Index. In all 3576 labels, like specimen. Complete, Price 5s.

VIOLACEÆ.
Viola lactea, Sm.
Cream-coloured Violet.—*Heaths, E. England.*
Loc.
Tem.　　|　　|　　Col.

The Preparation and Mounting of Microscopic Objects.

By THOMAS DAVIES. Fcp. cloth, price 2s. 6d.

This manual comprises all the most approved methods of mounting, together with the result of the Author's experience and that of many of his friends in every department of Microscopic Manipulation; and as it is intended to assist the beginner as well as the advanced student, the very rudiments of the art have not been omitted.

"Nothing is more difficult to those who handle a microscope for the first time than to get their objects in a fit state for exhibition and preparation. They will, therefore, feel greatly indebted to Mr. Davies for a little book on 'The Preparation and Mounting of Microscopic Objects.' It is clear, full, and practical; and it soon reveals to the careful student the valuable fact that a great deal may be done with very simple appliances. We recommend it to young microscopists as a book which supplies a felt deficiency."—*Guardian.*

Half-Hours with the Microscope.

By EDWIN LANKESTER, M.D., F.R.S. Illustrated by 250 Drawings from Nature by TUFFEN WEST.

CONTENTS.

Half an hour on Structure.
Half an hour in the Garden.
Half an hour in the Country.
Half an hour at the Pond-side.
Half an hour at the Sea-side.
Half an hour In-doors.

Appendix.—The Preparation and Mounting of Objects.

Third Edition, much enlarged, with full description of the various parts of the Instrument, price 2s. 6d. plain; 4s. coloured.

"The beautiful little volume before us cannot be otherwise than welcome. It is, in fact, a very complete manual for the amateur microscopist.... The 'Half-Hours' are filled with clear and agreeable descriptions, whilst eight plates, executed with the most beautiful minuteness and sharpness, exhibit no less than 250 objects with the utmost attainable distinctness."—*Critic.*

Hardwicke's Science-Gossip:

A Monthly Medium of Interchange and Correspondence for Students and Lovers of Nature. Monthly, 4d.; Annual Volume, in cloth, price 5s. See pages 56 to 64.

At Home in the Wilderness:

Being Adventures and Experiences in Uncivilized Regions, in which it is shown where and when to encamp; how to equip and manage a train of pack mules; break, gear, and saddle wild horses; cross streams, build log shanties, trenail a raft, dig out a canoe or build it with bark or hide, manage dog sleighs, tramp on snow shoes, &c. By J. KEAST LORD, late of the British North American Boundary Commission. Crown 8vo. cloth, price 6s.

In the Plain and on the Mountain.

A Guide for Pedestrians and Mountain Tourists in the Plain and on the Mountain. By CHARLES BONER, author of "Chamois Hunting in Bavaria," "Forest Creatures," &c. With illustrations of dress requisites, &c. Fcp. 8vo. price 2s.

"A little book which compresses into a very small space a great deal of good advice."—*Pall Mall Gazette.*

"We recommend Mr. Boner's book to all travellers, either on mountain or plain."—*Athenæum.*

The Book of Knots.

Illustrated by 172 Examples, showing the manner of making every Knot, Tie, and Splice. Price 2s. 6d.

"It is an honourable characteristic of our literature that it contains numerous admirable and complete treatises on many special subjects. Mr. Robert Hardwicke, publisher of many pleasant and useful works, has sent forth a 'Book of Knots,' by Tom Bowling, illustrated with one hundred and seventy-two diagrams, showing the manner of making every knot, tie, and splice, for the moderate price of half a crown, or nearly six knots a penny."—*All the Year Round*, May 23, 1868.

Old Bones; or, Notes for Young Naturalists.

By the Rev. W. S. SYMONDS, Rector of Pendock, Author of "Stones of the Valley," &c. With References to the Typical Specimens in the British Museum. Second Edition, much improved and enlarged, fully illustrated, fcp. 8vo. price 2s. 6d.

"The plan pursued by Mr. Symonds is a very simple one. He adopts the classification of Professsor Owen, and carries the young naturalist from family to family, beginning with man and ending with the lowest fishes, making his own remarks as he goes on. We recommend these notes. The volume is neatly got up, and deserves a sale amongst the class for whom it is intended."—*Athenæum.*

Notes on the Geology of North Shropshire.

By CHARLOTTE EYTON. Now ready, fcp. 8vo. cloth, price 3s. 6d.

Geological Stories;

A Series of Autobiographies in Chronological Order—Being the Autobiography of—A Piece of Granite; A Piece of Quartz; A Piece of Slate; A Piece of Limestone; A Piece of Sandstone; A Piece of Coal; A Piece of Rock-Salt; A Piece of Jet; A Piece of Chalk; A Piece of Purbeck Marble; A Lump of Clay; A Piece of Lignite; The Crags; A Boulder; A Gravel Pit. By J. E. TAYLOR, F.G.S., Author of "Half-hours at the Seaside," &c. Large fcap. 8vo., fully illustrated, price 7s. 6d.

The Handy-book of Rock Names.

By G. H. KINAHAN, of the Irish Geological Survey, fcap. 8vo., price 4s.

On the Being and Attributes of the Godhead,

As evidenced by Creation. By DAVID NELSON, M.D., Edinburgh, Author of "The Principles of Health and Disease," &c. Demy 8vo., price 10s. 6d.

The Applications of Geology to the Arts and Manufactures.

Six Lectures delivered before the Society of Arts. By Professor D. T. ANSTEAD, M.A., F.R.S. Fcap. 8vo. cloth, illustrated, price 4s.

I. On the Formation of Natural Soils by Derivation from Rocks, and on the Improvement of Soils by the admixture of Minerals.
II. On Natural and Artificial Springs, and on the various sources of Water Supply for Towns and Cities, in connection with the Geological Structure of the Vicinity.
III. On Mineral Materials used for the Purposes of Construction: Plastic and Incoherent Materials (Clays and Sands).
IV. On Mineral Materials (continued): Building Stones and Slates, and their Relative Value under given Circumstances of Exposure, and on Methods of Quarrying.
V. On Stratified Deposits of Minerals, as Coal and Iron Ore, usually obtained by Mining Operations, and on Mining Methods for such Deposits.
VI. On Metalliferous Veins for Lodes and their Contents, and on the Extraction of Metalliferous Minerals from Lodes.

"The science of geology largely engages the attention of the public; but persons are frequently deterred from the study of it by the dry and tedious style of writers dedicating their talent to its exposition. Professor Ansted has supplied a book which meets the public want.... The chapters relating to agricultural geology have an especial interest for farmers; but the whole are deeply interesting and worthy of recommendation."—*News of the World.*

"Professor Ansted takes in hand a subject of some difficulty, but of universal importance. In this he describes the different kinds of mineral veins, their contents, and the methods adopted to extract those contents, and thus fitly concludes his labours in elucidation of the practical application of geological knowledge to economic purposes, especially in connection with the arts."—*Daily News.*

Transactions of the Victoria Institute

Or, Philosophical Society of Great Britain and Ireland. Twenty parts, at 3s. 6d. to 5s. each. Lists on application.

Metamorphoses of Man and Animals.

Describing the changes which Mammals, Batrachians, Insects, Myriapods, Crustaceæ, Annelids, and Zoophytes undergo whilst in the egg; also the series of Metamorphoses which these beings are subject to in after-life. Alternate Generation, Parthenogenesis, and general Reproduction treated *in extenso*. With Notes, giving references to the works of Naturalists who have written upon the subject. By A. DE QUATREFAGES. Translated by HENRY LAWSON, M.D. Crown 8vo. cloth, price 6s.

"We have already said enough to show that the essay which Dr. Lawson has introduced to us in an English garb is one which marks a new era in the history of Embryology, and which presents to both general and scientific readers information which has been hitherto confined to the realms of dusty periodicals o all languages. 'The Metamorphoses' is a work which tends to elevate the science of Biology, and deserves the attention of all classes of cultivated readers."—*London Review.*

Science and Practice in Farm Cultivation.

By Professor BUCKMAN, F.L.S., F.G.S.

CONTENTS:

1. How to Grow Good Root Crops.
2. How to Grow Good Grass Crops.
3. How to Grow Good Clover Crops.
4. How to Grow Good Corn Crops
5. How to Grow Good Hedges.
6. How to Grow Good Timber.
7. How to Grow Good Orchards.

Fully Illustrated. Complete in One Volume, cloth, 7s. 6d.

The Stream of Life on our Globe.

Its Archives, Traditions, and Laws, as revealed by Modern Discoveries in Geology and Palæontology. A Sketch in Untechnical Language of the Beginning and Growth of Life, and the Physiological Laws which govern its progress and operations. By J. L MILTON, M.R.C.S. Second Edition, crown 8vo. cloth, pp. 624 price 6s.

CONTENTS:

The Beginning of Life.
England long, long ago.
The First Dwellers on Earth.
The First Builders.
The First Wanderers.
The First Colonists of Sacred History.
The First Language.
The First Alphabet.
The Battle of Life.

Glance at the Laws of Life.
Life in the Blood.
Life in the Nerves.
Life of a Giant.
Life of Men of Genius.
Influence of Smoking on Life and Race.
Life in the Stars and Planets; or Coloured Stars and their Inhabitants, &c. &c.

"A very agreeably-written record of some of the newest and most remarkable discoveries in geology, language, and physiology. The language is always un technical and picturesque. It has the merit of inspiring interest in subjects often treated in a manner to repel the ordinary reader."—*Lancet.*

Country Cottages.

A series of Designs for an Improved Class of Dwellings for Agricultural Labourers. By JOHN VINCENT, Architect. With numerous Plans, Elevations, &c. New Edition, folio, fully illustrated, price 12s.

Photographic Optics:

Including the Description of Lenses and Enlarging Apparatus. By D. VAN MONCKHOVEN, Ph.D. Crown 8vo. cloth, with more than 200 illustrations, price 7s. 6d.

BOOK I.—PHOTOGRAPHIC LENSES.

Preliminary Ideas.	Spherical Aberrations.
Refraction of Light.	Description of Photographic Objectives.
Chromatics.	
Lenses.	Employment of Photographic Objectives.
Aberrations.	

BOOK II.—APPARATUS FOR ENLARGEMENTS.

On the Negative intended for Enlargement.	Setting-up Enlarging Apparatus in Winter.
Description of Apparatus for Enlargements.	Setting-up of Movable Dyalitic Apparatus.
Theory of Formation of Enlarged Image in Woodward's Apparatus.	Application of Heliostat to Enlarging Apparatus.
Imperfections of Woodward's Apparatus.	Setting-up of Heliostat with Enlarging Apparatus.
The Dyalitic Apparatus.	Parallel Solar-Light Apparatus of Bertsch.
Description and Setting up of Dyalitic Apparatus, and of all Apparatus for Enlargements.	Indirect Enlargement by the Sun or by Diffused Light.
Management of Enlarging Apparatus.	Application of Artificial Light to Enlarging Apparatus.

Half-Hours at the Seaside:

Or, Recreations with Marine Objects. By J. E. TAYLOR, F.G.S., Author of "Geological Stories," &c. Small 8vo. with about 150 Illustrations, price 4s. plain; 6s. coloured.

CONTENTS:

Half an Hour with the Waves.	Half an Hour with Sea Anemones
Half an Hour with Preparations.	Half an Hour with Sea Mats and Squirts.
Half an Hour with Seaweeds.	
Half an Hour with Sponges.	Half an Hour with Sea Urchins and Starfish.
Half an Hour with Seaworms.	
Half an Hour with Corallines.	Half an Hour with Shell Fish.
Half an Hour with the Jelly Fish.	Half an Hour with Crustaceæ.

Half-Hours with the Telescope:

Being a Popular Guide to the Use of the Telescope as a means of Amusement and Instruction. Adapted to inexpensive Instruments. By R. A. PROCTOR, B.A., F.R.A.S. Fcap. 8vo. cloth, with Illustrations on Stone and Wood, price 2s. 6d.

CONTENTS:

Half an Hour on Structure of Instrument.	Half an Hour with Boötes, Scorpio, Ophiuchus, &c.
Half an Hour with Orion, Lepus, Taurus, &c.	Half an Hour with Andromeda, Cygnus, &c.
Half an Hour with Lyra, Hercules, Corvus, Crater, &c.	Half-hours with the Planets. Half-hours with the Sun, Moon, &c.

"It is crammed with starry plates on wood and stone, and among the celestial phenomena described or figured, by far the larger number may be profitably examined with small telescopes, and none are beyond the range of a good three-inch achromatic. The work also treats of the construction of telescopes, the nature and use of star maps and other subjects connected with the requirements of amateurs.... The book is full of 'useful and interesting information,' and will form a valuable companion to the various admirable handbooks for which Mr. Hardwicke enjoys so good a reputation."—*Illustrated Times.*

Half-hours with the Stars:

A Plain and Easy Guide to the Knowledge of the Constellations, showing, in 12 Maps, the Position of the Principal Star-Groups Night after Night throughout the Year, with Introduction and a separate Explanation of each Map. True for every Year. By RICHARD A. PROCTOR, B.A., F.R.A.S., late Scholar of St. John's College, Cambridge, and Mathematical Scholar of King's College, London ; Author of "Saturn and its System," "Half-hours with the Telescope," "The Handbook of the Stars," "Sun-views of the Earth," &c. &c. Second Edition, demy 4to. price 5s.

"Nothing so well calculated to give a rapid and thorough knowledge of the position of the stars in the firmament has ever been designed or published hitherto. Mr. Proctor's 'Half-hours with the Stars' will become a text-book in all schools, and an invaluable aid to all teachers of the young."—*Weekly Times.*

The Astronomical Observer:

A Handbook for the Observatory and the Common Telescope. By W. A. DARBY, M.A., F.R.A.S., Rector of St. Luke's, Manchester. Embraces 965 Nebulæ, Clusters, and Double Stars. Royal 8vo. cloth, price 7s. 6d.

"I think the design of the work has been well carried out. The catalogue will no doubt be very acceptable to the amateur observer desirous of obtaining a knowledge of practical astronomy, and it will also be useful in the library of the regular observatory."—*From the Earl of Rosse, K.P., F.R.S., &c.*

Chamber and Cage Birds.

Their Management, Habits, Diseases, Breeding, and Methods of Teaching them. Translated from the last German Edition of Dr. Bichstein's Chamber Birds, by W. E. SHUCKARD. New Edition, by GEO. J. BARNESBY, Judge of Show Birds, Derby. Price 3s. 6d.

Science and Commerce.

Their Influence on our Manufactures. A Series of Statistical Essays and Lectures describing the Progressive Discoveries of Science, the Advance of British Commerce, and the Activity of our Principal Manufactures in the Nineteenth Century. By P. L. SIMMONDS, Editor of *The Journal of Applied Science*, Honorary and Corresponding Member of Various Foreign and Colonial Societies. 600 pp. Fcap. 8vo. price 6s.

CONTENTS.

Obligations of Commerce to Science and the Vegetable Products Imported.
Mineral and Animal Substances entering into Commerce.
The Industrial and Manufacturing Uses of Shells.
The Progress of Science in the 19th Century.
The Cotton Manufacture:
  Production and Consumption of Cotton.
  Our Cotton Supplies.
  The Cotton Trade.
The Woollen Manufacture:
  The Wools of Commerce.
  Wool and the Woollen Trade.
  Our Wool Supplies.
  Colonial Wools.
  The British Woollen Manufactures.
    A—The Worsted Manufacture.
    B—The Woollen Manufacture.
    C—Shoddy Fabrics.
Chief Mineral and Vegetable Products of Commerce.

Silk Trade and Manufacture:
  On Silk Cultivation and Supply in India.
  Production of Silk.
Iron Manufacture.
Progress of our Mineral Industries.
Glass Manufacture.
Earthenware and Pottery.
Dyes and Colouring Stuffs:
  Dyes obtained from the Animal Kingdom.
  Mineral Colours and Dyes.
  Vegetable Colours and Dyes.
The Manufacture of, and Trade in, Precious Metals and Fancy Articles.
The Grocery and Allied Trades:
  The Commerce in Groceries.
  The Leather Trade and Tanning Substances:
  The Leather Manufacture.
New Paper-making Materials and Progress of the Paper Manufacture.
On Nuts: their Produce and Uses.

The Cattle Plague.

With Official Reports of the International Veterinary Congresses, held in Hamburg, 1863, and in Vienna, 1865. By JOHN GAMGEE, Demy 8vo. 860 pp. price 21s.

Horse Warranty.

A Plain and Comprehensive Guide to the Various Points to be noted, showing which are essential and which are unimportant. With Forms of Warranty. By PETER HOWDEN, V.S. Fcap. 8vo. price 3s. 6d.

Graceful Riding.

A Pocket Manual for Equestrians. By S. C. WAITE. With Illustrations. Fcap. 8vo. cloth, price 2s. 6d.

"In the school, on the road, on the course, or across country, this little book will be invaluable; and we heartily recommend it."—*Morning Post*.

One Hundred Double Acrostics.

A New Year's Gift. Edited by "Myself." 16mo. cloth, price 2s. 6d.

Hardwicke's Elementary Books. Price Twopence each.

Under the above title is presented to the public a complete Library of Elementary Works adapted for the use of the people.

Thousands of people at present skilled as handicraftsmen, and as workers in the various arts and sciences of life, plod on from day to day, with some vague notion that they can improve their own particular calling; but it generally ends in nothing, because they are ignorant of the first principles of those laws which regulate the things in which they are engaged. It is a rare occurrence to meet with workmen who know anything beyond what they picked up in their apprenticeship, or obtained by imitating others more skilled than themselves.

It will be the aim of HARDWICKE'S ELEMENTARY BOOKS to teach these first principles.

No labour or expense will be spared to make the information of a thoroughly reliable character; and where advisable, a free use of authentic illustrations will be brought to bear.

"Hardwicke's series of 'Elementary Books of Science' at present includes Optics, Hydraulics, Hydrostatics, Geography, Chemistry, Mechanics, and the parts vary in price from twopence to sixpence. Each part is a very admirable epitome of the subject it treats, and there is more reliable information in any one of these little pamphlets of a few pence than there is in many a costly volume. The woodcuts are in outline, or only slightly shaded, and their subjects are exceedingly well selected."—*London Review*.

NOW READY.

Mechanics.
Fully illustrated by nearly 100 cuts. Two parts, 2d. each; complete, 4d.; bound in cloth, 6d.

Hydrostatics.
Fully illustrated. Complete, 2d.; in cloth, 4d.

Chemistry.
Three parts, 2d. each; complete, 6d.; cloth, 8d.

Hydraulics.
Fully illustrated. Complete, 2d.; cloth, 4d.

Optics.
Fully illustrated. Complete, 4d.; cloth, 6d.

Pneumatics.
Fully illustrated, 2d.

Other Works of a similar character are in preparation, and will shortly be announced.

Mackenzie's Educational Books.
Intended for Schools or Self-instruction.

ALREADY PUBLISHED.

Mackenzie's Tables.
Commercial, Arithmetical, Miscellaneous, and Artificers'. Calculations in Bricklaying, Carpentry, Lathing, Masonry, Paperhanging, Paving, Painting, Plastering, Slating, Tiling, Wellsinking, Digging, &c. &c. Fractions and Decimals. Forms of Receipts and Bills. Calculations on Man, Steam, Railways, Power, Light, Wind, &c. Language and Alphabets. Calendar of the Church. Scripture Money, Principal Foreign Moneys and Measures. Geographical and Astronomical Tables, &c. &c. Complete, price 2d.; cloth, 6d.

Murray's English Grammar.
Complete, word for word with the Shilling Editions. Price 2d.; cloth, 4d.

Mavor's Spelling.
With numerous Cuts. Price 4d.; or, two parts, 2d. each.

Walkinghame's Arithmetic.
Same as the Half-Crown Edition. Price 4d.; or, two parts, 2d each; cloth, 6d.

Short-Hand.
With Phrases and Exercises, to gain facility in the use of all the characters, by which perfection may soon be attained. Complete price 2d.

Phrenology.
Explained and Exemplified. Complete, price 2d.

Bookkeeping.
By Single Entry, with explanations of Subsidiary Books, being a useful system for the Wholesale and Retail Shopkeeper. Complete, price 2d.

Elements of Arithmetic.

From the French of M. C. BRIOT. Translated by J. SPEAR, Esq. Crown 8vo. cloth, price 4s.

"The little book before us is a translation of a French school arithmetic, and we notice it especially in reference to the metric system, which is clearly explained in its pages, and put foward in the introduction as one of the prominent features of the work."—*Morning Star.*

Mangnall's Questions, Complete, 1s.

The Cheap Edition of this valuable School Book is now ready. It has been carefully revised and brought up to the present time. It is well printed and strongly bound.

"Published in a compact form, neatly bound, and being condensed without being abridged, comes before us in a greatly improved form. Few books contain so much information in so small a space."—*Portsmouth Guardian.*

Method for Teaching Plain Needlework in Schools.

By a LADY. Price 2s. 6d. Illustrated by Diagrams and Samplars

This useful method is based on *steps* which are *gradual*, *well defined*, and clea to the *perception*. It is calculated to insure the improvement of each individua child, and while it offers the necessary instruction to the less talented pupil, i enables the more clever one to attain the highest degree of perfection.

Oral Training Lessons in Natural Science and General Knowledge.

Embracing the subjects of Astronomy, Anatomy, Physiology Chemistry, Mathematical Geography, Natural Philosophy, th Arts, History, Development of Words, &c., intended for Teacher of Public Schools, and also for Private Instruction. B H. BARNARD, Principal of Lincoln's School, Minneapolis. Crow 8vo. price 2s.

Education and Employment of the Blind.

What it has been, is, and ought to be. By T. R. ARMITAG. M.D. Demy 8vo. cloth, price 2s. 6d.

Institutions and Charities for the Blind

In the United Kingdom (a Guide to), together with Lists of Bool and Appliances for their use; a Catalogue of Books publishe upon the subject of the Blind; and a List of Foreign Institution &c. By MANSFIELD TURNER and WILLIAM HARRIS. Demy 8v price 3s.

Dedicated by Express Permission to H.R.H. the PRINCE of WALES.

The County Families;

Or, Royal Manual of the Titled and Untitled Aristocracy of the Three Kingdoms. It contains a complete Peerage, Baronetage, Knightage, and Dictionary of the Landed Commoners of England, Scotland, Wales, and Ireland, and gives a brief notice of the Descent, Birth, Marriage, Education, and Appointments of each Person, his Heir Apparent or Presumptive, a Record of the Offices which he has held, together with his Town Address and Country Residences. By EDWARD WALFORD, M.A., late Scholar of Balliol College, Oxford. 1,200 pages, 11,000 families. Published Annually. Price £2. 10s.

"What would the gossips of old have given for a book which opened to them the recesses of every County Family in the Three Kingdoms—we will not say every recess, for here and there we observe what may be termed 'The Blue Beard family cupboard,' omissions which are not made evidently without very good cause! There are, for example, some awkward blanks of parentage to be found. Very many have no fathers; at all events none such as they cared to name. In some places the particulars of marriages are omitted, possibly with prudence. This work, however, will serve other purposes besides those of mere curiosity, envy, or malice. It is just the book for the lady of the house to have at hand when making up the county dinner, as it gives exactly that information which punctilious and particular people are so desirous of obtaining—the exact standing of every person in the county. To the business man, *The County Families* stands in the place of directory and biographical dictionary. The fund of information it affords respecting the upper ten thousand must give it a place in the lawyer's library; and to the money-lender, who is so interested in finding out the difference between a gentleman and a 'gent,' between heirs-at-law and younger sons, Mr. Walford has been a real benefactor. In this splendid volume he has managed to meet an universal want, one which cannot fail to be felt by the lady in her drawing-room, the peer in his library, the tradesman in his counting-house, and the gentleman in his club."—*Times.*

"It possesses advantages which no other work of the kind that we know of has offered hitherto. Containing all that is to be found in others, it furnishes information respecting families of distinction which are not to be found in the latter. It will prove to be invaluable in the library and drawing-room."—*Spectator.*

"To produce such a work in the perfection which characterizes 'County Families' must have been an almost Herculean task. It is sufficient for us to say that accuracy even in the minutest details appears to have been the aim of Mr. Walford, and the errors are so few and slight that they may readily be passed over."—*Weekly Register.*

By the same Author, published annually,

The Shilling Peerage,
The Shilling Baronetage,
The Shilling Knightage, and
The Shilling House of Commons,

Giving the Birth, Accession, and Marriage of each Personage, his Heir Apparent or Presumptive, Family Name, Political Bias and Patronage ; as also a brief Notice of the Offices which he has hitherto held, his Town Address and Country Residences.

The Complete Peerage, Baronetage, Knightage, and House of Commons.

In One Volume. Half-bound, with coloured edges, for marking the different divisions. Price 5s.

How to Address Titled People.

Companion to the Writing-Desk; or, How to Address, Begin, and End Letters to Titled and Official Personages, together with Tables of Precedence, copious List of Abbreviations, Rules for Punctuation, and other useful information. New Edition. Royal 32mo. price 1s.

"A word, and more than a word of praise is due. Full information on every subject of importance to correspondents is afforded in it. The instructions on official points will not fail to be of importance to many persons unable to obtain the proper information from even much larger works of the same kind."—*Court Journal.*

"This is one of the most useful little books we have for a long time seen."—*Era.*

The Royal Guide to London Charities.

Showing, in Alphabetical Order, the Name, Date of Foundation, Address, Object, Annual Income, Number of People benefited, Mode of Application to and Chief Officers of every Institution in London. By HERBERT FRY. Published Annually, price 1s. 6d.

Peter Schlemihl.

From the German of ADALBERT VON CHAMISSO. Translated by Sir JOHN BOWRING, LL.D., &c. Crown 8vo. cloth, with Illustrations by George Cruikshank, price 2s. 6d.; the Illustrations on India paper, price 5s.

Hardwicke's Shilling Handy-Book of London.

An Easy and Comprehensive Guide to Everything worth Seeing and Hearing. Royal 32mo., cloth, price 1s.

CONTENTS:

Bazaars.
Ball-rooms.
Cathedrals.
Dining-rooms.
Exhibitions.
Mansions of Nobility.
Markets.
Money-order Offices.
Monuments and Statues.
Museums.
Music-halls and Concert-rooms.
Out-door Amusements.

Omnibuses.
Palaces.
Parks.
Passport Offices.
Picture Galleries' Regulations.
Popular Entertainments.
Police-courts.
Prisons.
Railway Stations.
Steamboats.
Theatres.
Telegraph Offices, &c.

Shooting Simplified.
A Concise Treatise on Guns and Shooting. By JAMES DALZIELL DOUGALL. Second Edition, Re-written and and Enlarged, with a Special Chapter on Breech-Loaders. Fcap. 8vo. cloth, price 6s.

Wines and other Fermented Liquors.
From the Earliest Ages to the Present Time. Dedicated to all Consumers in the United Kingdom. By JAMES RICHMOND SHEEN. Fcap. 8vo. cloth, price 5s.

The Common Sense of the Water Cure.
A Popular Description of Life and Treatment at a Hydropathic Establishment. By Captain J. H. LUKIS, late of the 61st Regiment, and the North Durham Militia. Crown 8vo. price 5s.

"We have seldom read a more amusing book than this. Capt. Lukis is a clever well-bred gentleman, who has found the greatest pleasure of his life in the strict discipline of hydropathy, and in this volume he chats pleasantly about his own experiences, and puts in a very popular form the chief arguments in favour of the water cure. There is not a dull page in the book."—*Morning Herald.*

Special Therapeutics.
An Investigation into the Treatment of Acute and Chronic Disease by the Application of Water, the Hot-Air Bath, and Inhalation. By J. C. LORY MARSH, M.D., M.R.C.P. Crown 8vo. price 3s. 6d.

Bathing: How to do it, When to do it, and Where to do it.
By EDGAR SHEPPARD, M.D., Medical Superintendent of the Male Department, Colney Hatch Asylum. Third Edition. 8vo. sewed, price 1s.

Plain and Practical Medical Precepts.
Second Edition, revised and much enlarged. By ALFRED FLEISCHMAN, M.R.C.S. On a large sheet, price 4d.

The Home Nurse.
A Manual for the Sick Room. By ESTHER LE HARDY. Second Edition, fcap. 8vo. cloth, price 2s. 6d.

Air.	Practical Duties.	Setons, Issues.
Cleanliness.	Moral Duties.	Blisters and Plaisters.
Dress.	The Patient.	Chamber of Death.
Diet and Cookery.	Visitors.	&c. &c.

" In our notice of the first edition we expressed our approbation of the manner in which she had performed her task; and we are gratified to see that the useful lessons she inculcated both in regard to nursing and medical attendance have met with such general approval as to require another edition of her unpretending but really valuable volume."—*Lancet.*

On Teething of Infants.
Its prevalent Errors, Neglects and Dangers; their influence on the Health, and as causes of death of Children. Including the dangers of Teething Powders, Soothing Syrups, &c. By HENRY HANKS, L.R.C.P. Edinburgh, M.R.C.S. England, &c. Illustrated by Cases. Fcap. 8vo. cloth, 3s. 6d.

The Domestic Management of Infants and Children.
In Health and Sickness. By S. BARKER, M.D., Brighton. 8vo· price 5s. Also,

The Diet of Infancy and Childhood.
By S. BARKER, M.D. Demy 8vo. sewed, price 1s.

The Gastric Regions and the Victualling Department.
By AN OLD MILITIA SURGEON. The whole outward and inward man, from the crown of his head to the corns on his little toes, all tell the sad tale of the Gastric Regions' Wrongs. Crown 8vo. cloth, price 2s. 6d.

"This is a most useful, and by no means a dull or heavy book. . . . The old Militia Surgeon gives some most useful advice, in a pleasant practical manner, respecting different varieties of food and their effects upon the system."—*Observer*

The Foot and its Covering.
With Dr. Camper's work on "The Best Form of Shoe," translated from the German. By JAMES DOWIE. New Edition. Fcap. 8vo. cloth, illustrated, price 2s. 6d.

Auvergne:
Its Thermo-Mineral Springs, Climate, and Scenery. A new Salutary Resort for Invalids. By ROBERT CROSS, M.D. With three Tinted Lithographs, price 4s.

Sketches of Ceylon.
Sketches of the Inhabitants, Animal Life, and Vegetation, in the Lowlands and High Mountains of Ceylon, as well as the Submarine Scenery near the Coast, taken in a Diving Bell. By Baron EUGENE DE RANSONNETT. With 26 large Chromo-lithographs, taken from Life by the Author. Folio, £2. 10s.

First Help in Accidents.
Being a Surgical Guide in the absence or before the arrival of Medical Assistance. For the use of the Public, especially for Members of both Military and Naval Services, Volunteers, and Travellers. By CHARLES H. SCHAIBLE, M.D., Ph.D., Royal Military Academy, Woolwich. Fully illustrated.

First Help in Accidents (*continued.*)

Bites.	Choking.	Hanging.
Bleeding.	Cold.	Poisoning.
Broken Bones.	Dislocations.	Scalds.
Bruises.	Drowning.	Sprains.
Burns.	Exhaustion.	Suffocation.

And other Accidents where instant aid is needful. In sup. royal 32mo. cloth, price 2s. 6d.

"A most useful and interesting little book, which is besides prettily got up, and contains some accurate and nice woodcuts. In 200 small pages we find, condensed, whatever can be done in case of accidents until the arrival of a professional helper; in fact, the work is a short but complete manual, and will prove of great service to military men, volunteers, and tourists. . . . The instructions are plain and to the point, and there is a welcome absence of technical terms."—*Spectator.*

A Manual of Popular Physiology:

Being an Attempt to Explain the Science of Life in Untechnical Language. By HENRY LAWSON, M.D., Co-Lecturer on Physiology and Histological Anatomy in St. Mary's Hospital Medical School. Fcap. 8vo. with 90 Illustrations, price 2s. 6d.

Man's Mechanism.	Digestion.	The Kidneys.
Life.	Respiration.	Nervous System.
Force.	Heat.	Organs of Sense.
Food.	The Skin.	&c. &c.

"Dr. Lawson has succeeded in rendering his manual amusing as well as instructive. All the great facts in human physiology are presented to the reader successively; and either for private reading or for classes, this manual will be found well adapted for initiating the uninformed into the mysteries of the structure and function of their own bodies."—*Athenæum.*

Dr. Lankester on Food.

A Course of Lectures Delivered at the South Kensington Museum. By E. LANKESTER, M.D., F.R.S., F.L.S. New Edition. Price 4s.

Dr. Lankester on the Uses of Animals

In Relation to the Industry of Man. By EDWIN LANKESTER, M.D., F.R.S. A Course of Six Lectures, delivered at the South Kensington Museum. Crown 8vo. pp. 350, cloth, fully illustrated, price 3s.

Silk.	Soap.	Insects.
Wool.	Waste.	Furs.
Leather.	Sponges and Corals.	Feathers, Horns,& Hair.
Bone.	Shell-fish.	Animal Perfumes.

"The information is presented in the most lucid, graceful, and entertaining manner."—*Economist.*

"Every one who peruses them will be grateful to the author. The history of those creatures whose products become through man's skill so useful to him is given with such charming feeling that the interest of the reader is attracted and enchained, whether he wills or no."—*Era.*

Practical Physiology: a School Manual of Health.
Being a Practical Guide to the Means of Securing Health and Life. Intended for the use of classes and general reading. By Dr. LANKESTER, F.R.S. Fifth Edition, fcap. 8vo. illustrated, price 2s. 6d.

"It is copiously illustrated. There is not a school of any kind for males or females, rich or poor, in which the book might not be used as a text-book; indeed, it ought to be as common as an English Grammar. Few persons are capable of forming an idea of the increase of human happiness and material prosperity which would follow a more general appreciation of the laws of health."—*Lancet*.

Domestic Medicines: their Uses and Doses.
In the absence of professional assistance, with Tables of Weights and Measures; the preparation of Beverages suitable for the Sick Room. Poisons and their Antidotes. Sixth Thousand, 32mo. cloth, price 1s.

The Changed Cross.
Words by L. P. W. Illuminated by K. K. Dedicated to the Memory of those blessed ones who having, "through much tribulation," finished their course with joy, now rest from their labours; and to those also who are still running with patience the course set before them, "Looking to Jesus." Square 16mo. elegantly printed, with illuminated crosses and border lines, price 6s.

Spiritualism and Animal Magnetism.
By Professor G. G. ZERFFI, Ph.D. Second Edition. Crown 8vo Price 1s. 6d.

"This is a very clever book. It disposes of Spiritualism pretty conclusively, and it gives expositions of Animal Magnetism in connection therewith that are, as far as they go, as convincing as such efforts well can be."—*Church Herald*.

"A few more such books as this excellent little treatise, and we shall have no more table-turning or spirit-rapping. Perhaps, before long, the force at which we now either ignorantly tremble, or with equal ignorance sneer, may be doing material work."—*Land and Water*.

"As far as Professor Zerffi renders a more rational account of spiritualism than those who are enthusiastically devoted to the study, the little book under notice is worthy attention."—*Bath Journal*.

Degrees and "Degrees;"
Or, the Traffic in Theological, Medical, and other "Diplomas" Exposed. By the Rev. HENRY BELCHER, M.A., Assistant Master in King's College School. Dedicated, by permission, to the Rev. Alfred Barry, D.D., D.C.L., Canon of Worcester, Principal of King's College, London. Demy 8vo. price 1s. 6d.

Notes on Conjectural Amendments of certain Passages in Shakespeare's Plays.
By P. A. DANIEL. Crown 8vo. price 3s. 6d.

A Statesman at Home.
A Dramatic Fragment. By G. T. LOWTH. Crown 8vo. limp

MEDICAL AND SURGICAL WORKS.

The Pocket Guide to the Pharmacopœia.
Being an Explanatory Classification of its Drugs, Preparations, and Compounds. All Essentials being comprised in a form and size adapted to the Practitioners' Note Book. To enable the busy Practitioner to do justice to his vocation, and avail himself of what no memory could otherwise retain, this little book has been compiled by one who has long felt the want of some such aid. Limp cloth for the pocket-book, 1s.

Observations on Sea-Sickness,
And on some of the Means of Preventing it. By Sir JAMES ALDERSON, M.D., D.C.L., F.R.S., Consulting Physician to St. Mary's Hospital. Crown 8vo. cloth, price 2s.

Surgical Appliances and Minor Operative Surgery.
By THOMAS ANNANDALE, F.R.C.S., Edinburgh. Fcap. 8vo. cloth, price 5s.

Abstract of Surgical Principles.
By THOMAS ANNANDALE, F.R.S. and F.R.C.S. (Edin.) price 1s. each.

I. Inflammation, Suppuration, Abscess, Sinus and Fistula, Mortification, Ulceration, and Ulcers, price 1s.
II. Tumours, or Morbid Growths, price 1s.
III. Dislocations, price 1s.
IV. Fractures, price 1s.

Chemistry.
By Professor BRANDE, D.C.L., F.R.S.L., and Professor ALFRED S. TAYLOR, M.D., F.R.S., F.R.C.P. Lond. Fcap 8vo. cloth, 900 pages, price 12s. 6d.

"For clearness of language, accuracy of description, extent of information, and freedom from the pedantry and mysticism of modern chemistry, no other textbook comes into competition with it. . . . The best guide to the study of chemistry yet given to the world."—*Lancet.*

"Conceived and worked out in the most sturdy common-sense method, this book gives, in the clearest and most summary method possible, all the facts and doctrines of chemistry, with more especial reference to the wants of the medical student."—*Medical Times.*

On the Cure of Clubfoot, without Cutting Tendons.
By RICHARD BARWELL, F.R.C.S., Surgeon and Lecturer on Anatomy, Charing Cross Hospital. With 28 Photographs. Second Edition, greatly enlarged, crown 8vo. 7s. 6d.

By the same Author,

A Treatise on Diseases of the Joints.
Illustrated by Engravings on Wood. Demy 8vo. cloth, price 12s.

Also,

Causes and Treatment of Spinal Curvature.
Crown 8vo. with Woodcuts, price 6s.

A Manual of the Operations of Surgery.
For the Use of Senior Students, House Surgeons, and Junior Practitioners. Illustrated. By JOSEPH BELL, F.R.C.S. Edinburgh, Lecturer on Surgery, Assistant-Surgeon, Clinical Ward, Royal Infirmary. Surgeon to the Eye Infirmary, Edinburgh. Fcp. 8vo. price 6s.

On Inhalation as a Means of Local Treatment of the Organs of Respiration by Atomised Fluids and Gases.
By HERMANN BEIGEL, M.D., L.R.C.P. 8vo. cloth, illustrated, price 6s.

On Surgical Diseases of Women.
By I. BAKER BROWN, F.R.C.S. (by Exam.), Surgeon to the London Surgical Home. Third Edition, revised and enlarged, demy 8vo. price 15s.

By the same Author,

On Ovarian Dropsy:
Its Nature, Diagnosis, and Treatment. The Result of Thirty Years' Experience. Post 8vo. cloth, price 5s.

Also,

On Scarlatina and its Treatment.
Second Edition, fcp. 8vo. cloth, price 2s. 6d.

The Ophthalmoscope:

Its Varieties and its Uses. Translated from the German of ZANDER by R. B. CARTER, F.R.C.S. Eng. (by Exam.). With Notes and Additions. 68 Woodcuts and 3 beautiful Coloured Chromo-lithographs. Royal 8vo., price 9s.

On the Arcus Senilis; or Fatty Degeneration of the Cornea.

By EDWIN CANTON, F.R.C.S., Surgeon to the Charing Cross Hospital. With numerous illustrations. 8vo. cloth, price 10s. 6d.

The Chemical Processes of the British Pharmacopœia,

And the Behaviour with Re-agents of their Products. By HENRY J. CHURCH, F.C.S. Fcp. 8vo. price 3s.

By the same Author,

Carbolic Acid as a Disinfectant.

And as a means of preventing the spread of the Cattle Disease. 8vo. sewed, price 1s.

Disinfection and the Prevention of Disease.

By HENRY BOLLMAN CONDY. 8vo. sewed, price 1s.

Also,

Air and Water, their Impurities and Purification.

8vo. cloth, price 1s.

On Joint Diseases.

Their Pathology, Diagnosis, and Treatment; including the nature and treatment of Deformities and Curvatures of the Spine. By the late HOLMES COOTE, F.R.C.S, 8vo. cloth, price 9s.

Varicocele.

(Clinical Lecture on) Delivered at the Lock Hospital. By WALTER COULSON, F.R.C.S., Surgeon to the Lock Hospital and St. Peter's Hospital. Crown 8vo. price 2s. 6d.

Parturition and its Difficulties.

With Clinical Illustrations and Statistics of 13,783 Deliveries. By J. HALL DAVIS, M.D., F.R.C.P., President of the Obstetrical Society. New Edition, revised and enlarged, crown 8vo. cloth, price 7s. 6d.

By the same Author,

Flexions of the Uterus.
With Notes of the Ovarian and other Diseases treated in the Female Ward, Middlesex Hospital, July to July, 1863-64. Price 2s. 6d.

Gleet:
Its Pathology and Treatment. With Memoir on the Treatment of Stricture of the Urethra by Subcutaneous Division. By HENRY DICK, B.A., M.D., Surgeon to the National Orthopædic Hospital. Second Edition, with Woodcuts, price 5s. 6d.

On the Safe Abolition of Pain in Labour and Surgical
Operations by Anæsthesia with Mixed Vapours. By ROBERT ELLIS, M.R.C.S., L.S.A. Crown 8vo. cloth, price 2s. 6d.

On Penetrating Wounds of the Chest.
Founded on Actual Observations in the Camp General Hospital before Sebastopol. By PATRICK FRASER, M.D. Demy 8vo. cloth, price 5s.

Treatment of Diseases of the Skin.
By WILLIAM FRAZER, M.D. Fcp. 8vo. price 3s.

On Diseases of the Kidney, and Dropsy.
By Dr. S. J. GOODFELLOW, M.D., F.R.C.P., Senior Physician to the Middlesex Hospital and Lecturer on Medicine at the Middlesex Hospital Medical School. Crown 8vo. price 7s. 6d. With illustrations from Nature by Tuffen West.

Tooth Extraction.
A Manual on the proper mode of extracting Teeth, compiled from the latest authorities and designed for the use of Students and Junior Practitioners. With a Table showing in parallel columns, the names of all the Teeth, the instruments required for their extraction, and the most approved methods of using them. By JOHN GORHAM, M.R.C.S., Fellow of the Physical Society of Guy's Hospital. Price 1s.

On Epidemic Cholera and Diarrhœa:
Their Prevention and Treatment by Sulphur. Third Edition. By JOHN GROVE, M.D., M.R.C.S., &c. Price 1s.

By the same Author,

Epidemics Examined and Explained.
8vo. cloth, price 5s.

Dr. Grove's System of Medical Book-Keeping.
Only three books required, and no posting necessary. The complete set, suitable to carry a large practice about three years, £4. 14s. 6d. Prospectuses and Examples gratis on application.

On Diseases of the Stomach:
By S. O. HABERSHON, M.D., F.R.C.P., Physician to and Lecturer on Materia Medica and Therapeutics at Guy's Hospital, &c. Crown 8vo. price 5s.

An Essay on some Varieties and Effects of Cancerous Disease of Bone.
By WILLIAM HICKMAN, M.B., F.R.C.S., Surgeon to the Western General Dispensary. 8vo. illustrated, 3s. 6d.

On the Use of Artificial Teeth.
In the Prevention of Indigestion, and the Cure of Diseases due to Imperfect Nutrition. By ROBERT T. HULME, M.R.C.S.E. Crown 8vo. cloth, price 2s. 6d.

On the Management of Labour in Contracted Pelvis.
A Thesis which obtained the first prize from the Faculty of Medicine of Paris. By WILLIAM H. JONES, M.D. Paris, M.R.C.S. Eng. Demy 8vo. cloth, price 4s.

Via Medica:
A Treatise on the Laws and Customs of the Medical Profession, in relation especially to Principals and Assistants, with Suggestions and Advice to Students on Preliminary Education. By J. BAXTER LANGLEY, M.R.C.S., F.L.S. Crown 8vo. cloth, price 3s.

Sciatica, Lumbago, and Brachialgia.
Their Nature and Treatment and their immediate relief and rapid Cure by Hypodermic Injection of Morphia. By HENRY LAWSON, M.D., Assistant Physician to St. Mary's Hospital, and Lecturer on Physiology in St. Mary's Hospital Medical School. Crown 8vo. 7s. 6d.

Portrait of Baron Liebig.
From a Negative taken expressly for the Publisher. Mounted on cardboard, price 10s. 6d.; in elegant gilt frame, price 21s.

Posological Tables:
Being a Tabular Arrangement of all the Medicines contained in the British Pharmacopœia; with Dose, Action, and Form of Administration, containing also Appendix on Poisons. By ALEXANDER MILNE, M.D. Price 1s. 6d.

Diseases of the Skin:
(The Pathology and Treatment of). By J. L. MILTON, Senior Surgeon to St. John's Hospital for Diseases of the Skin; corresponding Member of the New York Dermatological Society. Demy 8vo., cloth, with photographs, price, 10s. 6d.

On Shock after Surgical Operations and Injuries.
With Special Reference to Shock caused by Railway Accidents. By EDWIN MORRIS, M.D., F.R.C.S. Oxon. Crown 8vo. cloth, price 3s. 6d.

Text-Book of Skin Diseases.
By Dr. ISIDOR NEUMANN, Lecturer on Skin Diseases in the Imperial University of Vienna. Translated from the Second German Edition by special permission of the Author. By ALFRED PULLAR, M.D. and C.M. Edin., Fellow of the Royal Medical and Chirurgical Society, Physician to the East London Hospital for Children. Translation revised by the Author. Sixty Woodcuts, illustrating recent microscopic investigations. Royal 8vo. cloth, price 12s. 6d.

Osteology for Students:
A concise Description of the Human Skeleton, adapted for the use of Students in Medicine, accompanied by an Explanatory Atlas of Plates. By ARTHUR TREHERN NORTON, Assistant-Surgeon to St. Mary's Hospital, and Assistant-Lecturer and Demonstrator of Anatomy at the Medical School. 8vo. cloth, price 7s. 6d. 2 vols. complete.

Affections of the Throat and Larynx.
The classification, description, and statistics of 150 consecutive cases occurring in the Throat Department of St. Mary's Hospital. By ARTHUR TREHERN NORTON, F.R.C.S., Assistant Surgeon and Surgeon in charge of the Throat Department of St. Mary's Hospital, Lecturer on Anatomy in the Medical School. Demy 8vo. cloth, price 4s.

The Ward Manual;
Or, Index of Surgical Disease and Injury. For the Use of Students. By T. W. NUNN, F.R.C.S., Surgeon to the Middlesex Hospital. Limp cloth, price 3s.

Hygiene of Air and Water.
Being a Popular Account of the effects of the impurities of Air and Water; their detection and the modes of remedying them. By WILLIAM PROCTER, M.D., F.C.S., Surgeon to the York Dispensary. Crown 8vo. cloth, price 2s.

Handbook of the Sphygmograph.
Being a Guide to its Use in Clinical Research. By J. BURDON SANDERSON, M.D., F.R.S., Physician to the Hospital for Consumption, Assistant-Physician to Middlesex Hospital, and Joint-Lecturer on Physiology in the Middlesex Hospital Medical College. Crown 8vo. cloth, price 3s. 6d. Illustrated.

On Rupture, Inguinal, Crural, and Umbilical.
The Anatomy, Pathology, Diagnosis, Cause and Prevention; with New Methods of effecting a Radical and Permanent Cure. Embodying the Jacksonian Prize Essay for 1861. By JOHN WOOD, F.R.C.S. Eng. (Exam.) Demonstrator of Anatomy at King's College, London; Assistant Surgeon to King's College Hospital. With numerous Illustrations by Bagg. 8vo. cloth, 12s. 6d.

Lungs and Heart.
A Guide to the Physical Diagnosis of Diseases of the Lungs and Heart; together with an introduction to the Examination of the Urine. By JAMES SAWYER, M.B., Lond., &c. Resident Physician Queen's Hospital, Birmingham. Crown 8vo., price 6s.

Preservation of Sight.
Three Lectures. By DAVID SMITH, M.D., Member of the Royal College of Surgeon of England, Extra Academical Lecturer on the Eye, Glasgow, &c. Crown 8vo. price 3s. 6d.

Clinical Notes on Uterine Surgery:
With Special Reference to the Management of the Sterile Condition. By J. MARION SIMS, M.D., late Surgeon to the Woman's Hospital, New York. 8vo. fully illustrated, price £1. 1s.

On Lithotomy.

By WILLIAM F. TEEVAN, B.A., F.R.C.S., Surgeon to the Wes London Hospital and to St. Peter's Hospital, and Lecturer on Anatomy at Westminster Hospital. Demy 8vo. price 1s.

Trousseau's Clinical Medicine.

Translated and Edited, with Notes and Appendices, by the late P. VICTOR BAZIRE, M.D., London and Paris. Parts 2 and 3 4s. each; Vol. I. bound in cloth, price 14s.
Vol. II., III., and IV. Translated by JOHN ROSE CORMAC, M.D., Edin., F.R.S.E., Fellow of the Royal College of Physicians, Edinburgh, formerly Lecturer in the Medical School of Edinburgh, &c. Sewed, price 12s.; in cloth, price 14s. each.

A few copies of the 2nd and 3rd parts are still on sale. The work is being now brought out by the New Sydenham Society. All who have bought the early parts as published by Mr. Hardwicke can have the completion from him through their bookseller in the ordinary way.

Spinal Debility:

Its Prevention, Pathology, and Cure, in relation to Curvatures, Paralysis, Epilepsy, and various deformities. By EDWARD W. TUSON, F.R.C.S. Demy 8vo. cloth, price 5s. illustrated.

Plain Directions for dealing with an Insane Patient.

By J. M. WINN, M.D., M.R.C.P., Senior Physician to the St. George's and St. James's Dispensary, late Medical Superintendent of Sussex House Lunatic Asylum, &c. Fcap. 8vo. cloth, price 1s.

PUBLICATIONS of the ROYAL COLLEGE OF SURGEONS.

List on Application.

PUBLICATIONS OF THE RAY SOCIETY.

British Hemiptera-Heteroptera.
By J. W. Douglas and John Scott, £1. 10s.

Cirripedia Family.
By C. Darwin, Esq. 2 vols. £2. 2s.

British Freshwater Polyzoa.
By Professor Allman. £1. 11s. 6d.

Recent Foraminifera.
By Professor Williamson. £1. 11s. 6d.

Oceanic Hydrozoa.
By Professor Huxley. £1. 11s. 6d.

Organization of Trilobites.
From the German of Burmeister. 15s.

British Naked-eyed Pulmograde Medusæ.
By Professor E. Forbes. £1. 1s.

The Spiders of Great Britain and Ireland.
By J. Blackwall, F.L.S. 2 vols. £3. 13s. 6d.

Foraminifera.
By Dr. Carpenter. £1. 11s. 6d.

Fructification of Higher Cryptogamia and Coniferæ.
From the German of Hofmeister. £1. 5s. 6d.

Reptiles of India.
By Dr. Günther. £2. 2s.

British Spongiadæ.
By Dr. Bowerbank. 2 vols. £2. 11s.

Steenstrup on Alternation of Generations.
15s.

British Entomostracous Crustacea.
By Dr. BAIRD. £1. 1s.

British Angiocarpous Lichens.
By Rev. W. A. LEIGHTON. 10s. 6d.

British Nudibranchiate Mollusca.
A Monograph of the (with coloured drawings of every species). By Messrs. ALDER and HANCOCK.

Part I. Imp. 4to.	.. .. ..	£1	10	0
,, II. ,,	.. ..	1	10	0
,, III. ,,	.. ..	1	10	0
,, IV. ,,	.. ..	1	10	0
,, V. ,,	.. ..	1	10	0
,, VI. ,,	.. ..	1	1	0
,, VII. ,,	.. ..	1	1	0

Meyen's Geography of Plants.
Translated from the German by Miss MARGARET JOHNSTON. 8vo. pp. 422, 10s. 6d.

The Miscellaneous Works of Robert Brown.
2 vols. demy 8vo. Vol. 1. I. Geographico-Botanical; II. Structural and Physiological Memoirs, price £1. Vol. 2. Systematic Memoirs and Contributions to Systematic Works. £1. Atlas of Plates, imp. 4to. price £1.

Vegetable Teratology.
An Account of the Principal Deviations from the usual construction of Plants. By MAXWELL T. MASTERS, M.D. F.L.S. With numerous illustrations, by E. M. WILLIAMS. Demy 8vo. cloth, £1. 1s.

On Pterylography.
Translated from the German. Edited by PHILIP LUTLEY SCLATER, M.A. Ph.D. F.R.S. Imp. 4to. 10 plates, 16s.

Recent Memoirs on the Cetacea.
By Professors ESCHRICHT, REINHARDT, and LILLJEBORG. Edited by W. H. Flower, F.R.S., &c. Imp. 4to. pp. 312, 6 plates, £1. 4s

The Shoulder Girdle and the Sternum.
By W. K. PARKER, Esq., F.R.S. Imp. 4to. illustrated with 30 coloured plates, £1. 11s. 6d.

Hydroids.
A Monograph of the Gymnoblastic or Tubularian Hydroids. By GEO. JAMES ALLMAN, M.D., Dublin and Oxon. 2 volumes, with coloured plates. Imp. 4to. £3. 3s. 0d.

ROBERT HARDWICKE, 192, *Piccadilly.*

POPULAR SCIENCE REVIEW.

EDITED BY HENRY LAWSON, M.D.

THE *Popular Science Review* is the only Journal which takes up the intermediate position between the world of Science and the public generally. There are scientific journals abundant which represent each section of the scientific public properly so called; and there are literary journals which devote a certain portion of their space to scientific matters of general interest. But the only journal which is at once a sort of connecting link between the scientific world on the one side and the people—*i.e.* those who are beyond the limit of the scientific circle—on the other, is the *Popular Science Review*.

Thus it contains no original theories. Until a theory, in fact, has obtained the sanction of the purely scientific world, it holds no place in the pages of the *Popular Science Review*; but as soon as a fact is recognized as an unquestionable scientific truth, and long before it reaches the general public by any other channel, it takes its place in the pages of the *Popular Science Review*. Here it is laid out in the plainest but most exact language possible, and, if necessary, is illustrated fully,—not in that claptrap style which is adopted by magazines of a particular class, but in a clear, distinct, and yet modest fashion, plainly setting forth the aims and desires of the author.

The plan of the *Popular Science Review* is briefly as follows:—First come a number of articles illustrated by a variety of page-plates, and dealing with those questions in each branch of science which have been most recently the subject of communications to the Scientific Societies. These are invariably done by men who are at once the most eminent and representative in the several branches of science to which they belong: we refer to the names of Contributors as a proof of this.

Leaving the Original Articles, we come next to the Reviews. These are short, but as a rule they are to the point; and they generally contain notices of every scientific book which has been published during the quarter.

Finally, we come to the Scientific Summary. In this we find a very perfect record of the principal work done at home and abroad in the following various branches:—Archæology, Botany, Geology, Palæontology, Chemistry, Mineralogy, Microscopy, Metallurgy, Mining, Physics, Photography, Zoology, Comparative Anatomy, and Medicine. This department is divided amongst a number of men, and generally it may be said to put the general reader *au courant* with the progress of science at home and abroad during the quarter which has elapsed previous to publication.

Thus, altogether, the Journal furnishes an amount of scientific information, in a popular and yet exact form, which cannot be found in any other English periodical.

Vols. I. to XI., Sets in Numbers, £5. 15s. 11 vols. cloth, £6. 14s. Half-morocco, £8. 8s. Cloth Covers for Binding, price 1s. 6d. each.

42   ROBERT HARDWICKE, 192, *Piccadilly.*

Popular Science Review, Vol. 1 contains :—

Corn. By Prof. J. Buckman, F.L.S., F.G.S., F.S.A., &c. Illustrated.
The Daisy. By Mrs. Lankester. Illustrated.
The Crown Animalcule. By P. H. Gosse, F.R.S. Illustrated.
The Lowest Forms of Life. By James Samuelson and Dr. Bruxton Hicks, F.L.S.
Iron and Steel. By R. Hunt, F.R.S.
Artificial Light. By Prof. Ansted, F.R.S.
The Breath of Life. By W. Crookes, F.R.S.
The West Coast of Equatorial Africa. With a Coloured Map.
The Great Comet of 1861. By J. Breen. Illustrated.
Caverns and their Contents. By Prof. Ansted, F.R.S.
The Lowest Forms of Life. Illustrated.
The Flower Animalcules. By P.H.Gosse, F.R.S. Illustrated.
Cotton. By Dr. Lankester, F.R.S. Illustrated.
Grass. By Prof. Buckman, F.L.S. Illustrated.
The Reflex Theory. By G. H. Lewes.
Solar Chemistry. By R. Hunt, F.R.S. With Coloured Diagram.
Optical Phenomena of the Atmosphere. By G. F. Chambers.
The Phosphorescence of the Sea. With a Plate.

The Sun and Solar Phenomena. With a Coloured Plate. By James Breen, F.R.A.S.
Light and Colour. With a Coloured Plate. By Robert Hunt, F.R.S.
The Application of Science to Electroplating. By G. Gore, F.R.S.
Artificial Precious Stones. By W. S. Howgrave.
The White Clover. By Mrs. Lankester. With two Plates.
The Human Heart. By Isaac Ashe, B.A., T.C.D.
The Great Exhibition of 1862.
The Britannia and Conway Tubular Bridges. With Page Plate. By W. C. Unwin, B.Sc.
Primitive Astronomy. With two Coloured Illustrations.
The Physics of a Sunbeam. With Coloured Plate. By R. Hunt, F.R.S.
The English California. By G. P. Bevan, F.G.S.
The Contents of Caverns (concluding Part). By D. T. Ansted, F.R.S.
The Microscope, with Directions for its Use. Illustrated with Woodcuts. By C. Collingwood, M.B., F.L.S.
The Builder Animalcules. With a Page Plate. By P. H. Gosse, F.R.S.
The Common Truffle. With a Page Plate. By Jabez Hogg, M.R.C.S., &c.

Popular Science Review, Vol. 2 contains :—

The British Oak. Two Page Illustrations. By Prof. Buckman, F.L.S., F.G.S., &c.
Tubifex Rivulorum, the Red Worm of our Rivers. With a Coloured Plate. By Edwin Ray Lankester.
Anæsthetics. By Dr. T. L. Phipson, M.B., F.C.S., &c.
Barmouth and its Scientific Attractions. With a Tinted Illustration.
The Electro-plating Process (concluding Part). With an Illustration. By George Gore.
Notes of the Exhibition of 1862.
The Mistletoe and Parasitic Plants. By Mrs. Lankester. Illustrated.
The Winter Life of Plants. By Harland Coultas.
The Vinegar Eel. By Jabez Hogg, F.L.S. With Tinted Plate.
The Eye of the Ox and its Microscopical Structure. By E. Beckitt Truman. With Tinted Plate.

Mars. By James Breen, F.R.A.S. With Coloured Plate.
Our Fresh-water Polyzoa. By Rev. W. Houghton, M.A., F.L.S. Two Page Plates.
The African Lion in its Native Haunts. By Jules Gerard.
The Human Skin. By Isaac Ashe, M.B. With Page Plate.
The Furniture Woods of Commerce. By P. L. Simmonds.
Fossil Birds. By S. J. Mackie, F.G.S. Illustrated.
History of a Beech-tree. By Harland Coultas.
The Antiquity of Man. By J. R. Leifchild, M.A.
On the Physical Geography of the Ionian Islands. By Prof. D. T. Ansted, M.A., F.R.S.
The Telescope. By James Breen, F.R.S.
Our British Jungermaniæ.

Rotifera or Wheel Animalcules. Part IV. The Flexible Creepers (Notommatina). By Philip Henry Gosse, F.R.S. Illustrated.

On the Morphology of Vinca minor (small Periwinkle). By James Buckman, F.L.S., F.G.S.
Colour Blindness. By Jabez Hogg, F.L.S. Illustrated.

PopularScience Review, Vol. 3 contains:—

Photographic Printing and Engraving, with a Photograph of a Complete Page of the *Times* Newspaper printed from Stone. By William Crookes, F.R.S.
Fresh Air. By E. Lankester, M.D., F.R.S.
Microscopic Fungi Parasitic on Living Plants. By M. C. Cooke. Coloured Illustrations.
On the Physical Geography of the Ionian Islands. By Prof. D. T. Ansted, M.A., F.R.S.
The Metropolitan Main Drainage Works, with a Map of the Works. By S. J. Mackie, F.G.S.
Diseased Pork, and Microscopic Worms in Man. By John Gamgee. Illustrated.
Bodily Work and Waste. By Francis T. Bond, M.D., B.A. (Lond.), F.C.S.
The Railway Tunnel through the Alps. With Tinted Illustration.
Greek Fire: its Ancient and Modern History. By B. W. Richardson, M.A., M.D.
Notes on Earthquakes. By Rev. W. S. Symonds.

On Printing Telegraphs. With Coloured Fac-simile Illustration. By R. S. Culley.
Herrings and Herring Fishing. By the Editor.
On Proper Clothing. By E. Lankester, M.D., F.R.S.
On the Absorption and Radiation of Heat. With two Page Illustrations. By H. Debus, Ph.D., F.R.S.
Botanical Exercises. By Rev. G. Henslow, M.A., F.L.S.
Pre-historic Dwellings. With Tinted Illustrations. By G. E. Roberts, F.A.S.L.
The Aniline Dyes. With Page of Silk Illustrations. By Dr. T. L. Phipson, F.C.S.
On the Action of Manures. By Baron Liebig.
The Old Red Sandstone Fishes of England. By E. Ray Lankester. Illustrated.
Oysters and Oyster Culture. By the Editor. Illustrated.
The Pneumatic Dispatch. By S. J. Mackie, F.G.S. Illustrated.
Thermometry. By G. F. Chambers.

Popular Science Review, Vol. 4 contains:—

Extraordinary Ships. By S. J. Mackie, F.G.S. Illustrated.
Good Food. By Edwin Lankester, M.D., F.R.S.
On the Origin of Local Floras of Great Britain and Ireland. By Harland Coultas. Illustarted.
Metamorphism. By Prof. D. T. Ansted, M.A., F.R.S., &c.
Morphological Peculiarities of Linaria spuria. By E. S. Magrath. Illustrated.
The Metric System. By James Spear. Illustrated.
Recent Investigations into the Natural History of the Red Coral. By the Editor. Illustrated.
What is a Stimulant? By Francis E. Anstie, M.D.
The Source of Heat in the Sun. By Robert Hunt, F.R.S. Illustrated.
Soup for Children. By Baron Liebig.
Vegetables, Fruit, and Water as Sources of Intestinal Worms. By T. Spencer Cobbold, M.D., F.R.S. Illustrated.
The Anatomy and Physiology of the Foraminifera. By W. C. Williamson, F.R.S. Illustrated.

Waves of Heat and Waves of Death. By B. W. Richardson, M.A., M.D., F.R.S.
Origin of our Kitchen Garden Plants. By Harland Coultas.
The Geology of Mineral Springs. By Francis T. Bond, M.D., B.A., F.C.S. Illustrated.
Train-signalling in Theory and Practice. By Charles V. Walker, F.R.S., F.R.A.S.
Extract of Meat. By Charles Boner.
On Sponges. By Robert Patterson, F.R.S.
The Physical Phenomena of other Worlds. By Robert Hunt, F.R.S.
Migration of European Birds. By A. Leith Adams, M.A., M.B., F.G.S., F.L.S.
On the Difficulties in Identifying many of the Lower Kinds of Algæ. By J. Braxton Hicks, M.D., F.R.S.
On the Oldest known Fossil (Eozoon Canadense) of the Laurentian Rocks of Canada; its Place, Structure, and Significance. By Prof. T. Rupert Jones, F.G.S.

Balloon Ascents and their Scientific Importance. By James Glaisher, F.R.S. Illustrated.
On the Highest Magnifying Powers and their Uses. By Lionel S. Beale, M.B., F.R.S.
Darwin's Observations on the Physiology of the Process of Fertilization in Plants. By M. C. Cooke. Illustrated.
What is a Tonic? By Edward Divers, M.D., F.C.S.
The Eruption of Etna. By S. J. Mackie, F.G.S.
Inside the Eye: the Ophthalmoscope and its Uses. By Ernest Hart, Ophthalmic Surgeon and Lecturer on Ophthalmic Surgery and Medicine at St. Mary's Hospital Medical School. Illustrated.
Is the Whitebait a Distinct Species? By the Editor. Illustrated.
On the Means of Communicating between Guards and Passengers on Railways. By Thomas Symes Prideaux.
Atlantic Telegraphy. By Robert Hunt, F.R.S. Illustrated.
On Pure Water. By Edwin Lankester, M.D., F.R.S.
On the Nature and Diagnostic Value of Raphides and Other Plant Crystals. By Prof. Gulliver, F.R.S.
On Lake Basins. By Prof. D. T. Ansted, M.A., F.R.S. Illustrated.
Epidemics Past and Present: their Origin and Distribution.
On the Microscopic Anatomy of an Insect Larva (Corethra plumicornis). By E. Ray Lankester. Illustrated.
The Moon. By James Breen, F.R.A.S. Illustrated.
Photography, and some of its Applications. By the Editor.

Popular Science Review, Vol. 5 contains:—

Insects Injurious to the Turnip Crops. By Rev. W. Houghton, M.A., F.L.S.
Coffee. By Baron Liebig.
Australia and Europe formerly One Continent. By Berthold Seemann, Ph.D., F.L.S., V.P.A.S.
On Ozone in Relation to Health and Disease. By B. W. Richardson, M.A., F.R.S.
Glaciers and Ice. By W. F. Barrett. Illustrated.
On the Phenomena of Motion, Sensitiveness in Climbing Plants. By Rev. G. Henslow, M.A., F.L.S. With Page Illustration.
On the Construction and Use of the Spectrum Microscope. By H. C. Sorby, F.R.S.
On the Volvox Globator. By J. Braxton Hicks, M.D. (Lond.), F.R.S.
Engraving with a Sunbeam. Woodbury's Relief-printing. By J. Traill Taylor. Illustrated.
Entozoon-like Bodies in the Muscles of Animals destroyed by Cattle Plague. By Lionel S. Beale, M.B., F.R.S.
Our House Spiders. By J. Blackwall, F.L.S.
Raised Beaches, and their Origin. By Edward Hull, B.A., F.G.S.
On Milk, and its Adulteration. By Augustus Voelcher, Ph.D., F.C.S., &c.
The Amœba: its Structure, Development, and Habits. By Prof. W. C. Williamson, F.R.S.
On the Solfatara and Fumaroles in the Neighbourhood of Naples. By Prof. D. T. Ansted, M.A., F.R.S.
The Graphotype Process. By the Editor. Illustrated.
Hydræ or Fresh-water Polypes. By Rev. W. Houghton, M.A., F.L.S.
How to Work with the Telescope. Part I. By R. A. Proctor, B.A., F.R.A.S.
On the Exhaustion of our Coal. By Leonard Lemoran, M.E.
On Hybridization among Plants. By Rev. G. Henslow, M.A., F.L.S.
On the Light-emitting Apparatus of the Glowworm. By Henry Fripp, M.D.
Sun-force and Earth-force. By Dr. Richardson.
The Eruption of Santorin. By Prof. D. T. Ansted, F.R.S.
On the Movements of the Diatomaceæ. By E. Ray Lankester.
Aerolites. By Townshend M. Hall, F.G.S.
On the Electrical Principles of the Atlantic Telegraph. By Prof. G. C. Foster.
The Bone Caverns of Gibraltar, Malta, and Sicily. By A. Leith Adams, A.M., M.B., F.G.S., &c.
Notes on Lophopus crystallinus. By J. Josselyn Ranson and T. Graham Ponton.
Genesis or Parthenogenesis? By H. E. Fripp, M.D.
Modern Views of Denudation. By Edward Hull, B.A., F.G.S.
How to Work with the Telescope. Part II. By Richard A. Proctor, B.A., F.R.A.S.

Popular Science Review, Vol. 6 contains:—

On Growth of some of the Algæ. Illustrated by a few of the Common kinds. By J. Braxton Hicks, M.D., F.R.S.
The Geology of Sinai. By Rev. E. W. Holland, M.A.
The Planet Mars in January, 1867. By Richard A. Proctor, B.A., F.R.A.S. Illustrated.
On Water-filters. By Edward Divers, M.D., F.C.S. Illustrated.
Our Fresh-water Entomostraca, Shell Insects, or Water-Fleas. By W. Baird, M.D., F.L.S. Illustrated.
How to Photograph Microscopic Objects. By Edward T. Wilson, M.B. Oxon. Illustrated.
Recent Discoveries in Insect Embryogeny. By Henry Fripp, M.D. Illustrated.
On the Struggle for Existence among Plants. By J. D. Hooker, M.D., F.R.S.
How to Study Meteorology. By G. T. Chambers, F.R.A.S. Illustrated.
On Sensitive Flames. By W. F. Barrett. Illustrated.
Paraffin Lamps and their Dangers. By John Attfield, Ph.D., F.C.S.
Venus's Flower-basket (Euplectella). By Dr. J. E. Gray, F.R.S. Illustrated.

An Attempt to Approximate the Date of the Flint Flakes of Devon and Cornwall. By Spence Bate, F.R.S. Illustrated.
Jupiter without his Satellites. By Richard A. Proctor, B.A., F.R.A.S. Illustrated.
Fitzroy Weather Forecasts. By G. F. Chambers, F.R.A.S. Illustrated.
On Life Insurance and Vital Statistics. By W. Hardwicke, M.D.
New Electro-Magnetic Machines. By S. J. Mackie. Illustrated.
The Botany of a Coal-mine. By Wm. Carruthers, F.L.S. Illustrated.
The Microscope in Geology. By David Forbes, F.R.S. Illustrated.
Why the Leaves Fall. By Maxwell T. Masters, M.D., F.L.S.
A Message from the Stars. By Robert Hunt, F.R.S.
On the Planariæ of our Ponds and Streams. By E. Ray Lankester. Illustrated.
Ventilation and Ventilators. By the Editor. Illustrated.
Physics of the Brain. By B. W. Richardson, M.A., M.D., F.R.S.

Popular Science Review, Vol. 7 contains:—

The Common Fresh-water Sponge (Spongilla fluviatilis). By Prof. W. C. Williamson, F.R.S. Illustrated.
The Hurricane, the Typhoon, and the Tornado. By Prof. D. T. Ansted, F.R.S.
Sensitive Plants. By Maxwell T. Masters, M.D., F.L.S. Illustrated.
The former range of the Reindeer in Europe. By W. Boyd Dawkins, M.A., F.R.S.
The Science of a Snow-flake. By Robert Hunt, F.R.S. Illustrated.
The Food of Plants. By Prof. Church, M.A., F.C.S.
The Gems and Precious Stones of Great Britain. By J. Morris, F.G.S. Illustrated.
Is the Fresh-water Sponge (Spongilla) an Animal? By John Hogg, M.A., F.R.S.
How to Dissect a Flower. By M. C. Cooke. Illustrated.
The Polariscope and how to Work with it. By C. Hockin, M.A. Illustrated.
Free Nematoids. By H. Carlton Bastian, M.D., F.R.S. Illustrated.

On the Animals which are most nearly intermediate between Birds and Reptiles. By Prof. Huxley, LL.D., F.R.S. Illustrated.
The Study of Chemical Geology. By David Forbes, F.R.S., &c.
Plants known by their Pollen-grains and other Cells. By George Gulliver, F.R.S.
The Great Eclipse of August 17, 1868. By R. A. Proctor, B.A., F.R.A.S. Illustrated.
On the Range of the Mammoth. By W. Boyd Dawkins, M.A., F.R.S.
Caddis-worms and their Metamorphoses. By Rev. W. Houghton, M.A., F.L.S. Illustrated.
What is Wine? By August Dupré, Ph.D.
Iron Shields and Iron Forts. By S. J. Mackie, Assoc. Inst. C.E. Illustrated.
The Air or Swimming Bladder of Fishes. By Rev. W. Houghton, M.A., F.L.S.
How to make a Geological Section. By Prof. D. T. Ansted, M.A., F.R.S. Illustrated.
The Life of a Seed. By Maxwell T. Masters, M.D., F.L.S. Illustrated.

Popular Science Review, Vol. 8 contains:—

Flying Machines. By Fred. W. Breary, Hon. Sec. to Aëronautical Society.
The Compound Eye of Insects and Crustacea. By Henry Fripp, M.D. Illustrated.
True and False Flint Weapons. By N. Whitley, C.E.
The Planet Mars in February, 1869. By Richard A. Proctor, B.A., F.R.A.S. Illustrated.
On the Molecular Origin of Infusoria. By J. Hughes Bennett, M.D., F.R.S.E. Diagrams.
The Cuttle-fish. By St. George Mivart, F.Z.S. Illustrated.
The Nature of the Interior of the Earth. By David Forbes, F.R.S.
On the Use and Choice of Spectacles. By R. Brundell Carter, F.R.C.S. Diagrams.
The Use of the Spectroscope in Astronomical Observation. By Richard A. Proctor, B.A., F.R.A.S. Illustrated.
The British Lion. By W. Boyd Dawkins, M.A., F.R.S.
Passion-flowers. By Maxwell T. Masters, M.D., F.L.S. Illustrated.
The Natural Development of Bacteria in the Protolamic Parts of Various Plants. By M. A. Bechamp.
The Planet Saturn in July, 1869. By R. A. Proctor, B.A., F.R.A.S.

The Sertularian Zoophytes of our Shores. By Rev. T. Hincks, B.A. Illustrated.
Hydrogenium. By Robert Hunt, F.R.S.
The Structure and Affinities of the Seasquirts (Tunicata). By J. C. Galton, M.A., F.L.S. Illustrated.
The Fertilization of Salvia, and of some other Flowers. By William Ogle, M.D. Illustrated.
"In Articulo Mortis." By Benjamin W. Richardson, M.D., F.R.S.
Experimental Illustrations of the Modes of Determining the Composition of the Sun and other Heavenly Bodies by the Spectrum. By Wm. Allen Miller, M.D., D.C.L., V.P.R.S. Illustrated.
What is Bathybius? By Prof. Williamson, F.R.S.
Are there any Fixed Stars? By Richard A. Proctor, B.A., F.R.A.S. Illustrated.
Kent's Hole. By W. Boyd Dawkins, M.A., F.R.S.
The Lingering Admirers of Phrenology. By Prof. Cletand.
The Anatomy of a Mushroom. By M. C. Cooke. Illustrated.
The Chemistry of a Comet. By Edward Divers, M.D., F.C.S.

Popular Science Review, Vol. 9 contains:—

The Heat of the Moon. By J. Carpenter, F.R.A.S.
"Under Chloroform." By B. W. Richardson, M.D., F.R.S.
The Deposits of the Atlantic in Deep Water and their Relation to the White Chalk of Cretaceous Period. By Prof. D. T. Ansted, M.A., F.R.S.
What is Wine? By A. Dupré, Ph.D.
The Fertilization of Certain Plants (Didynamia). By W. Ogle, M.D., Oxon, F.R.C.P.
On some Interesting Points in the History of the Polyzoa. By Rev. Thomas Hincks, B.A.
The Structure of Rock Masses (Stratification Joints, Cleavage). By David Forbes, F.R.S., &c.
The Planet Jupiter 1669-70. By Rev. T. W. Webb, M.A.
Microscopic Test Objects under Parallel Light and Corrected Powers. By Rev. J. B. Reade, F.R.S., P.R.M.S.
The Fertilization of Various Flowers by Insects (Compositæ, Ericaceæ, &c.). By W. Ogle, M.D., F.R.C.P.
On the Sphærosira Volvox of Ehrenberg. By Prof. Williamson, F.R.S.

The Turret Ships *Monarch* and *Captain*. By S. J. Mackie, C.E. Illustrated.
The Usefulness of the Fifth in Music. By Rev. C. Hope Robertson, M.A.
The Anatomy of the River Mussel. By John C. Gatton, M.A., Oxon. F.L.S.
On a Simple Decimal System for England. By Royston Pigott, M.A., M.D., Cantab., M.R.C.P., F.C.P.S., F.R.A.S.
What Fills the Star Depths. By R. A. Proctor, B.A., F.R.A.S.
On the Apparatus employed in Deep Sea Explorations on Board H.M.S. *Porcupine* in the Summer of 1869. By W. Lant Carpenter, B.A., B.Sc.
The Geology of Mont Cenis Tunnel. By Prof. Ansted, F.R.S. With Map and Section.
Greenwich Time and its Telegraphic Distribution. By W. Ellis, F.R.A.S.
The Echinus, or Sea Urchin. By S. G. Mivart, F.R.S. Illustrated.
The Sun's Corona. By R. A. Proctor, B.A., F.R.A.S.
Machine Guns or Mitrailleuses. By S. J. Mackie, C E.

Popular Science Review, Vol. 10 contains :—

Hitting the Mark; or, Cannon-balls and their Striking Velocity. By G. W. Royston Pigott, M.A., M.D. Illustrated.
Natural Selection Insufficient to the Development of Man. By the Rev. George Buckle, M.A.
Polymorphic Fungi. By M. C. Cooke, M.A. Illustrated.
The Eclipse Expeditions. By R. A. Proctor, B.A., F.R.A.S.
Notes on Butterflies. By Rev. C. Hope Robertson, F.R.M.S. Illustrated.
On Sleep. By Dr. Richardson, F.R.S.
The Discophores or Large Medusæ. By Rev. T. Hincks, B.A. Illustrated.
The Issues of the Late Eclipse. By J. Carpenter, F.R.A.S.
Grafting; its Consequences and Effects. By Maxwell T. Masters, M.D., F.R.S. Illustrated.
Coal as a Reservoir of Power. By Robert Hunt, F.R.S.
The Plymouth Breakwater Port. By S. J. Mackie, C.E. Plate.
The Structure of Rock Masses (Foliation and Striation). By David Forbes, F.R.S. Illustrated.

South Africa and its Diamonds. By T. Rupert Jones, F.G.S.
British Bears and Wolves. By W. Boyd Dawkins. M.A., F.R.S.
The " Lotus " of the Ancients. By M. C. Cooke, M.A. Illustrated.
Greenland. By William Pengelly, F.R.S. F.G.S.
Observations on Jupiter in 1870-71. By Rev. T. W. Webb, M.A., F.R.A.S.
The International Exhibition at South Kensington. By S. J. Mackie, C.E. Illustrated by Heliotype Process.
How Fishes Breathe. By J. C. Galton, M.A., M.R.C.S., F.L.S. Illustrated.
Mr. Crooke's New Psychic Force. By J. P. Earwaker.
The Moss World. By R. Braithwaite, M.D., F.L.S. Illustrated.
Theory of a Nervous Ether. By Dr. Richardson, F.R.S.
On Pleistocene Climate and the Relation of the Pleistocene Mammalia to the Glacial Period. By W. Boyd Dawkins, M.A., F.R.S., F.G.S. Illustrated.
Star Streams and Star Sprays. By R. A. Proctor, B.A., F.R.A.S.

Popular Science Review, Vol. 11 contains:—

Mimicry in Plants. By A. W. Bennett, M.A.B., Sc. F.L.S. Illustrated.
Recent Microscopy. By Henry J. Slack, F.G.S., Sec. R.M.S.
Experimental Researches on the Contortion of Rocks. By L. C. Miall. Illustrated.
Psychic Force and Psychic Media. By J. P. Earwaker, Merton Col., Oxford.
Strange News about the Solar Prominences. By Richard A. Proctor, B.A. Illustrated.
Madder Dyes from Coal. By Edward Divers, M.D., F.C.S.
On the Structure of Camerated Shells. By H. Woodward, F.G.S., F.Z.S. Illustrated.
On the Temperature and the Movements of the Deep Sea. By Dr. W. B. Carpenter, F.R.S.
The Eclipse of last December. By R. A. Proctor, B.A., Sec. R.A.S.
The Lithofracteur. By S. J. Mackie, E.C.C. Illustrated.
The Physiological Position of Alcohol. By Dr. Richardson, F.R.S.
The Nature of Sponges. By Henry J. Slack, F.G.S. Illustrated.

On the Probable Existence of Coal-measures in the South-east of England. By Joseph Prestwich, F.R.S., F.G S. Illustrated.
Bud Variation. By Maxwell T. Masters M.D., F.R.S.
An Account of a Ganoid Fish from Queensland (Ceratodus). By Dr. Günther, F.R.S. Illustrated.
Greenwich Observatory. By James Carpenter, F.R.A.S. Illustrated.
The Recent Fossil Man. By J. Morris, F.G.S., Prof. of Geology in University College, London. Illustrated.
The Hydroid Medusæ. By Rev. Thomas Hincks, B.A., F.R.S. Illustrated.
The First Chapter of the Geological Record. By David Forbes, F.R.S., &c.
Electrical Signalling and the Siphon Recorder. By J. Munroe, Assistant to Sir W. Thomson.
Spontaneous Movements in Plants. By A. W. Bennett, M.A.B., Sec. F.L.S. Plate LXXXIX.
News from the Stars. By Richard A. Proctor, B.A., F.R.A.S.
Life Form of the Past and Present. By Henry Woodward, F.G.S., F.Z.S.

All Microscopists who care to know what is going on at home or abroad should take in—

THE

MONTHLY MICROSCOPICAL JOURNAL.

Transactions of the Royal Microscopical Society, and Record of Histological Research at Home and Abroad.

THIS Journal is devoted exclusively to the interests of Microscopical Science in the widest and most accurate sense of the term. It contains not only the proceedings of the Royal Microscopical Society, but also embraces communications from the leading Histologists of Great Britain, the Continent, and America, with a comprehensive *résumé* of the latest Foreign Inquiries, Critical Reviews and Short Notices of the more important works, Bibliographical Lists, and Descriptions of all new and improved forms of Microscopes and Miscroscopic Apparatus ; Correspondence on all matters of Histological Controversy ; and finally, a Department of "Notes and Queries," in which the student can put such questions as may elicit the special information he desires to obtain.

The Editor has also made arrangements for the publication of the most important Papers read before Local Associations. Contributions requiring illustration are accompanied by most carefully-drawn Plain or Coloured Plates, and the text is printed in clear and legible type, thus affording the Microscopist a readable Monthly Record of all that takes place in the branches of science specially interesting to him. By thus providing a journal at once thoroughly scientific, advanced, and comprehensive, and issued at such short intervals as to meet the requirements of active investigators, the Publisher hopes to receive the support of all workers with the Microscope, and the assistance and co-operation of all who desire to possess a periodical which creditably represents the labours of British and Foreign Histologists.

ROBERT HARDWICKE, 192, *Piccadilly.*

The Monthly Microscopical Journal,

Volume 1, *price* 10s. 6d., *392 Pages of Letter-press,* 17 *whole-page Plates, and numerous Woodcuts, contains :—*

Structure of Papillæ and Termination of Nerves in Muscle of Common Frog's Tongue. By Dr. Maddox. With Plate.

Relation of Microscopic Fungi to Cholera. By Dr. J. L. W. Thudichum.

Heliostat for Photo-Micrography. By Dr. Maddox. With plate.

A Modification of the Binocular Microscope. By M. Nachet. Illustrated.

Heliostat for Photo-Micrography. By Lieut.-Colonel J. J. Woodward, M.D., U.S. Army Medical Deparment. With Plate.

The Vital Functions of the Deep-Sea Protozoa. By Dr. G. C. Wallich.

The Formation of Blastoderm in Crustacea. By M. van Beneden, Brussels.

On the Classification and Arrangement of Microscopic Objects. By J. Murie, M.D., F.L.S.

Immersion Objectives and Test-Objects. By John Mayall, jun., F.R.M.S.

Notes on Mounting Animal Tissues. By H. Charlton Bastian, M.D., F.R.S.

Some Undescribed Rhizopods from North Atlantic Deposits. By G. C. Wallich, M.D., F.L.S.

On the Construction of Object-Glasses. By F. H. Wenham.

On the Organ of Hearing in Mollusks. By M. Lacaze-Duthiers.

On a New Infusoria. By J. G. Tatem, F.L.S.

The Composite Structure of Simple Leaves. By John Gorham, M.R.C.S.

The Construction of Object-Glasses for ths Microscope. By F. H. Wenham.

On Triarthra longiseta. By C. T. Hudson, LL.D.

On a New Growing Slide. By C. J. Muller.

Professor Owen's Views on Magnetic and Vital Forces. By Lionel S. Beale, M.B., F.R.S.

Scale-bearing Podurae. By S. J. McIntyre, F.R.M.S.

On the Fibres of the Crystalline Lens of Petromyzonini. By George Gulliver, F.R.S.

Two New Forms of Selenite Stages. By Frederick Blankley, F.R.M.S.

Researches on the Constitution and Development of te Ovarian Egg of the Sacculinæ. By M. J. Gerbe.

On the Simple Structure of Compound Leaves. By W. R. M'Nab, M.D. Edin.

On the Microscopical Structure of some Precious Stones. By H. C. Sorby, F.R.S., &c.

Construction of Object-Glasses. By F. H. Wenham.

On the Rhizopoda, Primordial Type of Animal Life. By G. C. Wallich, M.D., F.L.S., &c.

On the Red Blood Corpuscle of Oviparous Vertebrata. By William S. Savory, F.R.S.

A Small Zoophyte-trough. By W. P. Marshall.

Preparation of Rock Sections. By David Forbes, F.R.S., &c.

Markings on the Pleurosigma angulatum and Lepisma saccharina. By J. B. Dancer, F.R.A.S

Notes on Zoosperms of Crustacea. By Alfred Sanders, M. R. C. S., F.R.M.S.

Protoplasm and Living Matter. By Dr. Lionel S. Beale, F.R.S.

On some New Infusoria from the Victoria Docks. By Wm. S. Kent, F.R.M.S.

Parkeria and Loftusia. By Dr. Carpenter, V.P.R.S., and H. B. Brady, F.L.S.

The Microscope in Silkworm Cultivation. By M. Cornalia.

On the Proboscis of the Blow-fly. By W. T. Suffolk, F.R.M.S.

A New Universal Mounting and Dissecting Microscope. By W. P. Marshall, President of the Birmingham Natural History and Microscopical Society.

On Crystals enclosed in Blowpipe Beads. By H. C. Sorby, F.R.S.

On Free-swimming Amœbæ. By J. G. Tatem, Esq.

Action of Anæsthetics on the Blood Corpuscles. By J. H. McQuillen, M.D., D.D.S.

Note on the Blood-vessel System of the Retina of the Hedgehog. By J. W. Hulke, F.R.S.

A New Process of Preparing Specimens of Filamentous Algæ for the Microscope. By A. M. Edwards.

The Monthly Microscopical Journal,

Volume 2, *with* 344 *Pages of Letterpress,* 19 *whole-page Plates, price* 10s. 6d., *contains :—*

On the Rectal Papillæ of the Fly. By B. T. Lowne, M.R.C.S. With Plate.

On the Diatom Prism, and the True Form of Diatom Markings. By the Rev. J. B. Reade, M.A., F.R.S.

Observations on the Supposed Cholera Fungus. By the Rev. M. J. Berkeley, M.A., F.L.S.

On the Correlation of Microscopic Physiology and Microscopic Physics. By John Browning, F.R.A.S.

Notes on Hydatina lenta. By C. T. Hudson, LL.D. With plate.

Some Remarks on the Structure of Diatoms and Podura Scales. By F. H. Wenham.

Structure of the Adult Human Vitreous Humour. By David Smith, M.D., M.R.C.S.

On the Use of the Chloride of Gold in Microscopy. By Thomas Dwight, jun., M.D.

On a Simple Form of Micro-Spectroscope. By John Browning, F.R.A.S.

On the Structure and Affinities of some Exogenous Stems from the Coal-Measures. By W. C. Williamson, F.R.S. With plate.

On the Battledore Scales of Butterflies. By John Watson, Esq. With Three Plates.

On Methods of Microscopical Research. By Herr S. Stricker.

On the Construction of Object-Glasses for the Microscope. By F. H. Wenham.

Jottings from the Note-book of a Student of Heterogeny. By Metcalfe Johnson, M.R.C.S.

A Supposed Mammalian Tooth from the Coal-measures. By T. P. Barkas, F.G.S.

On Holtenia, a Genus of Vitreous Sponges. By Wyville Thompson, LL.D., F.R.S.

Icrospectroscopy. Results of Spectrum Analysis. By Jabez Hogg, F.L.S.

Memorandum of Spectroscopic Researches on the Chlorophyll of Various Plants. By the late William Bird Herapath, M.D., F.R.S.

Observations on Mucor Mucedo. By R. L. Maddox, M.D. With plate.

Floscularia coronetta, a New Species, with Observations on some Points in the Economy of the Genus. By Charles Cubitt, Assoc. Inst. C.E., F.R.M.S. With two plates.

On the Detection by the Microscope of Red and White Corpuscles in Blood-stains. By Joseph G. Richardson, M.D.

On the Staining of Microscopical Preparations. By Dr. W. R. M'Nab.

Some further Remarks on an Illumination for verifying the Structure of Diatoms, and other Minute Objects. By F. H. Wenham.

On the Rhizopodal Fauna of the Deep Sea. By William B. Carpenter, M.D., V.P.R.S.

On the Structure of the Stems of the Arborescent Lycopodiaceæ of the Coal-measures. By W. Carruthers, F.L.S. Illustrated.

On the Development of the Ovum of the Pike. By E. B. Truman. Illustrated.

On the Presence of Foraminifera in Mineral Veins. By Charles Moore.

On the Relations of the Ciliary Muscle to the Eye of Birds. By Henry Lawson, M.D. Illustrated.

Experiments on Spontaneous Generation. By Edward Parfitt, Curator of the Devon and Exeter Institution.

The Histology of the Eye. By John Whitaker Hulke, F.R.S., F.R.C.S., Assistant-Surgeon to the Middlesex Hospital, and Surgeon to the Royal London Ophthalmic Hospital.

On Collecting and Mounting Entomostraca. By J. G. Tatem.

Further Remarks on the Nineteen-Band Test-plate of Nobert, and on Immersion Lenses. By Col. Woodward.

On High Power Definition, with Illustrative Examples. By G. W. Royston Pigott, M.A., M.D.

My Experience in the Use of Microscopes. By Dr. H. Hagen.

Further Remarks on the Plumules of Battledore Scales of some of the Lepidoptera. By John Watson.

The Development of Organisms in Organic Infusions. By C. T. Staniland Wake, F.A.L.S.

The Monthly Microscopical Journal,

Volume 3, 334 Pages of *Letter-press*, 20 *whole-page Plates, price* 10s. 6d., *contains :—*

Structure of the Scales of Certain Insects of the Order Thysanura. By S. J. McIntire, F.R.M.S.
Organisms in Mineral Infusions. By C. Staniland Wake, F.A.L.S.
Markings on the Podura Scale. By G. Royston Pigott, M.D.
Cultivation, &c., of Microscopic Fungi. By R. L. Maddox, M.D.
Jottings by a Student of Heterogeny. By Metcalfe Johnson, M.R.C.S. No. 2.
The Mode of Examining the Microscopic Structure of Plants. By W. R. M'Nab, M.D. Edin.
On the Microscopical Examination of Milk under certain Conditions. By J. B. Dancer, F.R.A.S.
On the Stylet-Region of the Ommatoplean Proboscis. By W. C. McIntosh, M.D., F.R.S.E., F.L.S.
On a Method of Measuring the Position of Absorption-bands with a Micro-Spectroscope. By J. Browning, F.R.A.S., F.R.M S.
On an Undescribed Stage of Development of Tetrarhynchus corollatus. By A. Sanders, M.R.C.S., F.R.M.S.
On a New Instrument for Cutting Thin Sections of Wood. By M. Mouchet, Hon. F.R.M.S.
On the Calcareous Spicula of the Gorgonaceæ their Modification of Form, and the Importance of their Characters as a Basis for Generic and Specific Diagnosis. By W. S. Kent, F.L.S.
On Pollen; as an Aid in the Differentiation of Species. By C. Bailey, Esq.
On Professor Listing's Recent Optical Improvements in the Microscope. By Dr. H. Hagen.
On the Structure of the Stems of the Arborescent Lycopodiaceæ of the Coal-Measures. By W. Carruthers, F.L.S., F.G.C., Botanical Dept., Brit. Museum.
The Mode of Examining the Microscopic Structure of Plants. By W. R. McNab, M.D. Edin.
Description of some Peculiar Fish's Ova. By W. B. Carpenter, M.D.,

On the Shell Structure of Fusulina. By W. B. Carpenter, M.D., F.R.S.
On the Comparative Steadiness of the Ross and the Jackson Microscope-stands. By W. B. Carpenter, M.D., F.R.S.
A New Method of Using Darker's Films. By Edward Richards.
A New Tube-dwelling Stentor. By C. A. Barrett, M.R.C.S.
A Contribution to the Teratology of Infusoria. By J. G. Tatem.
The Polymorphic Character of the Products of Development of Monas Lens. By Metcalfe Johnson, M.R.C.S.E.
On the Reparation of the Spines of Echinida. By W. B. Carpenter, M.D., V.P.R.S.
On the Colouring Matters derived from the Decomposition of some Minute Organisms. By H. C. Sorby, F.R.S., &c.
Cercariæ, Parasitic on Lymnæa stagnalis, By Jabez Hogg, Hon. Sec., R.M.S.
Experimental Researches on the Anatomical and Functional Regeneration of the Spinal Cord. By M. M. Masius and Van Lair.
Observations on some Points in the Economy of Stephanoceros. By C. Cubitt, Assoc. Inst. C.E., F.R.M.S.
Notes on Diatomaceæ. By Prof. Arthur Mead Edwards.
The New Binocular Microscope. By Samuel Holmes.
Reminiscences of the Early Times of the Achromatic Microscope. By J. S. Bowerbank, LL.D., F.R.S.
On an Apparatus for Collecting Atmospheric Particles. By R. L. Maddox, M.D.
The Magnesium and Electric Light as applied to Photo-Micrography. By Brevet Lieut.-Colonel J. J. Woodward, U. S. Army.
Remarks on High Power Definition. By F. H. Wenham, R.M.S.
On a New Critical Standard Measure of the Perfection of High Power Definition, as afforded by Diatoms and Nobert's Lines. By Dr. Royston Pigott.

The Monthly Microscopical Journal,

Volume 4, 342 *Pages of Letter-press*, 14 *whole-page Cuts, price* 10s. 6d., *contains*:—

On Fungi and Fermentation. By J. Bell, F.C.S.
On the Origin of the Colouring Matter in Mr. Sheppard's Dichroic Fluid. By E. Ray Lankester, B.A., F.R.M.S.
Object-Glasses and their Definition. By F. H. Wenham.
On the Optical Advantages of Immersion Lenses and the Use of Deviation Tables for Optical Research. By Royston Pigott, M.A., M.D.
On Synchæta mordax. By C. T. Hudson, LL.D.
Notes on Diatomaceæ. By Prof. Arthur Mead Edwards.
On an Erecting Binocular Microscope. By J. W. Stephenson, F.R.A.S., F.R.M.S.
Further Remarks on the Oxycalcium Light, as applied to Photo-Micrography. By Brevet Lieut.-Col. J. J. Woodward.
Cursory Remarks on the Podura Scale, Lepidocyrtus curvicollis, and Degeeria domestica, or the Speckled Variety. By R. L. Maddox, M.D.
Description of a Simple Air Sieve. By Metcalfe Johnson, S.C.E.
The Microscopic Structure of the Human Liver. By Dr. H. D. Schmidt, of New Orleans.
Microscopic Examination of the Atmosphere. By Geo. Sigerson, M.D.
Microscopical Examination of Rocks and Minerals. By S. Allport, F.G.S.
On the Structure of the Pleurosigma angulatum and Pleurosigma quadratum. By John Anthony, M.D., Cantab.
Object-Glasses and their Definition. By F. H. Wenham.
The Ciliary Muscle and Crystalline Lens in Man. By J. W. Hulke, F.R.S.
On the Preparation of Specimens of Soundings for the Microscope. By Prof. A. M. Edwards.
Circulation of the Latex in the Laticiferous Vessels. By H. C. Perkins, M.D.
On a New Species of Parasite from the Tiger. By T. Graham Penton, F.Z.S.
On the Application of the Microscope to the Study of Rocks. By H. C. Sorby, F.R.S., &c.
On the Focal Length of Microscopic Objectives. By C. R. Cross.
The Patterns of Artificial Diatoms. By H. J. Slack, Sec. R.M.S.
Ancient Water-Fleas of the Ostracodous and Phyllopodous Tribes Bivalved Entomostraca. By Professor T. Rupert Jones, F.G.S.
On the Real Nature of Disease-Germs. By Lionel S. Beale, F.R.S.
On the Histology of Minute Blood-vessels. By Brevet Lieut.-Col. Woodward.
On the Formation of Microscopic in closed Cells. By A. W. Wills.
The Ciliary Muscle and Crystalline Lens in Man. By J. W. Hulke.
On the "Hexactinellidæ" or Hexradiate Spiculed Silicious Sponges taken in the "Norna" Expedition. By W. Saville Kent, F.Z.S.
On a Mode of Ascertaining the Structure of Scales of Thysanuradeae. By Joseph Beck, F.R.M.S.
On the Advancing Aplanatic Power of the Microscope and New Double Star and Image Tests. By G. W. Royston Pigott, M.A., M.D.
American Microscopes and their Merits. By C. Stodder.
On a New Anchoring Sponge, Dorvillia agariciformis. By W. Saville Kent, F.Z.S.
On Aplanatic Definition and Illumination, with Optical Illustrations. By G. Royston Pigott, M.A., M.D., &c.
On Selecting and Mounting Diatoms. By Cap. F. H. Lang.
On Certain Cattle-plague Organisms. By Boyd Moss, F.R.C.S.
Notes on New Infusoria. By J. G. Tatem.

ROBERT HARDWICKE, 192, *Piccadilly.*

The Monthly Microscopical Journal,

Volume 5, 286 *Pages of Letterpress,* 20 *whole-page Plates, contains:—*

Notes on Fluorescence v. Pseudo-dichroism. By the Late Rev. J. B. Reade, F.R.S., P.R.M.S.
Notes on the Minute Structure of the Scales of Certain Insects. By S. J. McIntire, F.R.M.S.
On an Optical Illusion Slide; Cracks in Silica Films. By Henry J. Slack, F.R.M.S.
Object-Glasses and their Definition. By F. H. Wenham.
On the Mounting of the Diatom Prism. By F. W. Griffin, Ph.D.
On Pterodina valvata: A New Species. By C. T. Hudson, LL.D.
Observations on the Use of the Aërocouiscope, or Air-Dust-collecting Apparatus. By R. L. Maddox, M.D.
On the Employment of Colloid Silica in the Preparation of Crystals for the Polariscope. By Henry J. Slack, F.G.S., Sec. R.M.S.
The Development of Phycocyan. By T. C. White, M.R.C.S.
The Anatomy of the Round Worm (Ascaris lumbricoides, Linn.). By B. T. Lowne, M.R.C.S., Eng.
On the History, Refractions, Definition, and Powers of Immersion Lenses and New Refractometers. By Royston Pigott, M.A., M.D.
On Microscopical Appliances. By Dr. Royston Pigott.
Recent Investigations into Minute Organisms. By H. J. Slack, Sec. R.M.S.
On a New Form of Binocular Eye-piece and Binocular Microscope for High Powers. By C. D. Ahrens.
On Crystalline Forms modified by Colloid Silica. By Henry James Slack, F.R.M.S.

Object-Glasses and their Definition. By F. H. Wenham, R.M.S.
Nobert's Nineteenth Brand and its Observers. By Charles Stodder.
On the Structure of the Podura Scale, and Certain other Test Objects by Photo-Micrography. By Lieut.-Col. D. M. Woodward.
Microscopical Examination of Water for Domestic Use. By James Bell, F.C.S.
On the Winter Habits of the Rotatoria. By C. Cubitt, F.R.M.S.
The Magnifying Power of the Microscope. By Count Castracane.
A Few Experiments bearing on Spontaneous Generation. By Metcalfe Johnson, M.R.C.S.E.
On the Mode of Working out the Morphology of the Skull. By W. Kitchen Parker, F.R.S., President R.M.S.
Linear Projection considered in its Application to the Delineation of Objects under Microscopic Observation. By C. Cubitt, F.R.M.S.
Optical Appearances of Cut Lines in Glass. By Henry J. Slack.
Transmutation of Form in Certain Protozoa. By Metcalfe Johnson, M.R.C.S.E.
Microscopical Examination of Two Minerals. By Professor A. M. Edwards.
Additional Observations concerning the Podura Scale. By Lieut. J. J. Woodward.
Remarks on the Scales of some of the Lepidoptera, as "Test Scale" of Lepidocyrtus curvicollis. By R. L. Maddox, M.D.
On the so-called Suckers of Dytiscus and the Puvilli of Insects. By B. T. Lowne, M.R.C.S.

The Monthly Microscopical Journal,

Volume 6, 302 *Pages of Letterpress and* 18 *whole-page Plates, price* 10s. 6d., *contains :—*

On Bog Mosses. By R. Braithwaite, M.D., F.L.S.
Structure of Podura Scales. By F. H. Wenham, Vice-president R.M.S.
On some New Parasites. By T. Graham Ponton, F.Z.S.
On some Improvements in the Spectrum Method of Detecting Blood. By H. C. Sorby, F.R.S., &c.
On the Cellular Structure of the Red Blood Corpuscle. By Joseph G. Richardson, M.D.
On the Use of Nobert's Plate. By Assistant-Surgeon J. J. Woodward, U. S. Army.
On the Employment of Damma in Microscopy. By Professor Arthur Mead Edwards, New York.
Experiments on Angular Aperture. R. B. Tolles.
Mycetoma: the Madura or Fungus-foot of India. By Jabez Hogg, Hon. Sec. R.M.S.
Diatomaceous Earth from the Lake of Valencia, Caracas. By A. Ernst, Esq., and H. J. Slack, F.G.S.
The Silicious Deposit in Pinulariæ. By Henry J. Slack, F.G.S., Sec. R.M.S,
Observations and Experiments with the Microscope on the Chemical Effects of Chloral Hydrate and other Agents, on the Blood. By Thomas Shearman Ralph, M.R.C.S., England.
Floscularia Cyclops; a New Species. By Charles Cubitt, F.R.M.S.
Mr. Tolles's Experiments on Angular Aperture." By F. H. Wenham.
On the Microscopical Structure and Composition of a Phònolite from the "Wolf Rock." By S. Allport, F.G.S., with Chemical Analysis by J. A. Phillips.
On Spore-cases in Coals. By J. W. Dawson, LL.D., F.R.S.
On a New Rotifer. By C. T. Hudson.
On the Examination of Mixed Colouring Matters with the Spectrum Microscope. By H. C. Sorby, F.R.S.
On Spectra formed by the passage of Polarized Light through Double-refracting Crystals seen with the Microscope. By Francis Deas, M.A.
Remarks on some Parasites found on the Head of a Bat. By R. L. Maddox, M.D.
Notes on the Resolution of Amphipleura pellucida by a Tolles Immersion ⅛th. By Assistant-Surgeon J. J. Woodward, U. S. Army.
Micro-ruling on Glass and Steel. By John F. Stanistreet, F.R.A.S., with Illustrative Remarks by Henry J. Slack.
The Fungoid Origin of Disease and Spontaneous Generation. By Jabez Hogg, Hon. Sec. R.M.S.
On an Improved Method of Photographing Histological Preparations by Sunlight. By J. J. Woodward.
Hæmatozoa in Blood of Ceylon Deer. By Boyd Moss, M.D.
Microscopical Fissures in the Masticating Surface of Molars and Bicuspids. By J. H. M'Quillen, M.D.
Transmutation of Form in Certain Protozoa. By Metcalfe Johnson.
On Gnats' Scales. By Jabez Hogg, Esq., Hon. Sec. R.M.S.
The Examination of Nobert's Nineteenth Band. By F. A. P. Barnard.
An Incident in the Life of a Chelifer. By S. J. McIntire, F.R.M.S.
On the Form and Use of the Facial Arches. By W. K. Parker, F.R.S., President R M.S.
On the Angular Aperture of Immersion Objectives. By Robt. B. Tolles, of Boston, U. S. A.
Notes on Pedalion Mira. By C. T. Hudson, LL.D.
Another Hint on Selecting and Mounting Diatoms. By Capt. F. H. Lang.
The Monads' Place in Nature. By Metcalfe Johnson, M.R.S.C.E.
Mapping with the Micro-Spectroscope, with the Bright-line Micrometer. By H. G. Bridge.
Some Remarks on a "Note on the Resolution of Amphipleura pellucida by a Tolles immersion ⅛th. By Assistant-Surgeon J. J. Woodward, U. S. Army.
Infusorial Circuit of Generations. By Theod. C. Hilgard.
Notes on Professor James Clark's Flagellate Infusoria, with Description of New Species. By W. Saville Kent, F.Z.S., F.R.M.S.
Instrument for Micro-ruling on Glass and Steel. By J. F. Stanistreet.
On the Conjugation of Amœba. By J. G. Tatem, Esq.
Crystallization of Metals by Electricity under the Microscope. By Philip Braham, Esq.
Infusorial Circuit of Generations. By Theod. C. Hilgard.
On the Connection of Nerves and Chromoblasts. By M. Geo. Punhett.

The Monthly Microscopical Journal,

Volume 7, 294 *Pages of Letterpress*, 20 *whole-page Plates, price* 10s. 6d., *contains :—*

Markings on Battledore Scales of some Lepidoptera. By John Anthony, M.D. Cantab., F.R.M.S.

The Nerves of Capillary Vessels and their Probable Action in Health and Disease. By Dr. Lionel S. Beale, F.R.S., F.R.C.P.

New Erecting Arrangement for Binocular Microscopes. By R. H. Ward, M.A., M.D.

On a New Micrometric Goniometer Eye-piece for the Microscope. By J. P. Southworth.

Action of Hydrofluoric Acid on Glass, viewed Microscopically. By H. F. Smith.

On the Relation of Nerves to Pigment and Other Cells or Elementary Parts. By Dr. Lionel S. Beale, F.R.C.P.

On the Structure of the Stems of the Arborescent Lycopodiaceæ of the Coal-Measures. By W. Carruthers, F.R.S.

On Bog Mosses. By R. Braithwaite, M.D., F.L.S.

The Advancing Powers of Microscopic Definition. By Dr. Royston Pigott, M.A., Cantab.

Microscopic Object-Glasses and their Power. By Edwin Bicknell.

Remarks on a Tolles Immersion $\frac{1}{10}$th. By Edwin Bicknell.

Maltwood's Finder supplemented. By W. K. Bridgman.

On a New Micro-Telescope. By Prof. R. H. Ward.

Mycetoma, the Fungus Disease of India. By Jabez Hogg, F.L.S.

The American Spongilla, a craspedote, flagellate Infusorian. By H. James Clarke, A.B., B.S.

Refractive Powers of Peculiar Objectives. By R. B. Tolles (U.S.).

On the Development of Vegetable Organisms within the Thorax of Living Birds. By Dr. James Murie, F.L.S., F.G.S., &c.

Remarks on the Finer Nerves of the Cornea. By Dr. E. Klein.

Note on the Resolution of Amphipleura pellucida by certain Objectives made by R. and J. Beck and by William Wales. By Dr. J. J. Woodward, U. S. Army.

Stephenson's Erecting Binocular. By J. W. Stephenson, F.R.A.S.

On a Presumed Phase of Actinophryan Life. By J. G. Tatem.

On the Various Phenomena exhibited by the Podura Test under Microscopic Resolving Powers. By G. W. Royston Pigott, M.A., M.D.

Researches on the First Stages of the Development of the Common Trout (Salmo-Fario). By Dr. E. Klein.

On the Classification and Arrangement of Microscopic Objects. By Dr. James Murie, F.L.S., F.G.S., &c.

On Bichromatic Vision. By J. W. Stephenson, F.R.A.S., Treas. R.M.S.

The Supposed Fungus on Coleus Leaves, and Notes on Podisoma fuscum and P. juniperi. By Henry J. Slack, F.G.S.

Optical Curiosities of Literature. By the Rev. S. Leslie Brakey, M.A.

On an Improved Reflex Illuminator for the Highest Powers of the Microscope. By F. H. Wenham.

On a Silvered Prism for the Successive Polarization of Light. By J. W. Stephenson, F.R.A.S., Treas. R.M.S.

Structure of Battledore Scales. By J. Anthony, M.D., F.R.M,S.

Beale's Nerve Researches. Dr. Beale. in Reply to Dr. Klein.

Crystallization of Metals by Electricity By Philip Braham.

On the Means of Distinguishing the Fibres of New Zealand Flax from those of Manilla or Sizal by the Microscope. By Captain Hutton.

ROBERT HARDWICKE, 192, *Piccadilly*.

CHEAP NATURAL HISTORY PERIODICAL.

Hardwicke's Science-Gossip:

An Illustrated Medium of Interchange and Gossip for Students and Lovers of Nature,

About Animals, Aquaria, Bees, Beetles, Birds, Butterflies, Ferns, Fish, Fossils, Lichens, Microscopes, Mosses, Reptiles, Rocks, Seaweeds, Wildflowers.

Edited by J. E. TAYLOR, F.G.S., Author of "Half-hours at the Seaside," "Geological Stories," &c.

"This is a very pleasant journal, that costs only fourpence a month, and from which the reader who is no naturalist ought to be able to pick up a good fourpenny-worth of pleasant information. It is conducted and contributed to by expert naturalists who are cheerful companions, as all good naturalists are; technical enough to make the general reader feel that they are in earnest, and are not insulting him by writing down to his comprehension, but natural enough and direct enough in their records of facts, their questioning and answering each other concerning curiosities of nature. The reader who buys for himself their monthly budget of notes and discussions upon pleasant points in natural history and science, will probably find his curiosity excited and his interest in the world about him taking the form of a little study of some branch of this sort of knowledge that has won his readiest attention. For when the study itself is so delightful, and the enthusiasm it excites so genuine and well-directed, these enthusiasms are contagious. The fault is not with itself, but with the public if this little magazine be not in favour with a very large circle of readers."— *Examiner.*

Hardwicke's Science-Gossip, Vol. 1. Selection from the Index.*

Amœba, the, 45.
Ants, 113, 116, 143, 179, 185, 234, 235, 239, 262, 263.
Aphides, Swarms of, 287.
Aquarian Difficulties, 154, 188, 213, 239.
Badger, the, 87.
Balance of Power, 193.
Bees, 34, 41, 93, 137, 143, 166, 167, 185, 214, 257, 263, 286, 287.
Bitten by a Viper, 131.
Black Cradle, 270.
Breeze-fly, the, 194.
Bromley, and What I found there, 246.
Cabbage Butterfly and its Metamorphoses, 30, 74.
Cat-fleas, 278.
Caterpillars, Brood of, 126, 168, 288.
Chapter from the Life of a Volvox, 244.
Circle of Life, 145.
Cleaning Diatomaceæ, 52.
Colour of Birds' Eggs, 39, 47, 87, 142, 231.
Common Things Unknown, 88.
Coral Reefs, 112, 220, 285.

Crickets, 42, 66, 84, 113, 128, 166.
Cui Bono? 25.
Dead Fly on the Window, 10.
Diatoms, 27, 52, 85, 95, 114, 140, 148, 163, 167, 237, 250.
Diet of Worms, 180, 214.
Duckweeds, 5, 258, 286.
Ferns, 20, 34, 37, 44, 66, 67, 93, 109, 114, 117, 187, 188, 190, 214, 291, 262, 284.
Fly in Pike-fishing, 280.
Fly Parasites, 93, 227.
Foot of a Fly, 253.
Four Years' Acquaintance with a Toad, 12.
Gathering Seaweeds, 173.
Gossamer Spiders, 151, 191, 213.
Gossip about Mansuckers, 49.
Green Drake-fly, 231.
Hair worm, more Notes on the, 197.
Hairs, Chapter on, 29.
Hermit Rooks, 226.
House Ants, 170, 239.
House Fly, the, 82.
Humming-bird Hawk-moth, 208.

---

* In making a selection from the Index to each volume] space precludes giving more than about one-tenth of each volume.

Imperfectly-developed Flowers, 103.
Independence, 241.
"In Memoriam," 265.
Insect-moulds, 133.
Intelligence of Starling, 13
Jelly Animalcule, 58.
Jelly-fishes, 248.
Keyhole Limpet and Parasite, 122.
Leaf Teachings, 52.
London Rocket, 149.
Lord Scarabæus, 98.
Microscopic Illumination, 130.
Mistletoe, 114, 273, 283.
Mounting Objects, 46, 65, 93, 94, 116, 163, 191.
Notes on the Hawthorn, 198.
Old Trees, 222.
Orchids, How to Grow, 124, 162.
Otter-shell, the, 79.
Pigment-cells, 106.
Piratical Gulls, 272.
Plantain, the, 232.
Plant Animals, 177.
Plea for Nettles, 275.
Polàrized Light, 224.
Polycystius, Popular History, 100.
Puzzle worth thinking about, 127.
Rock Whistler, the, 242.

Sea-Anemones, 40, 155, 158, 167, 188 190, 196, 213, 239, 260, 285, 286, 287.
Sea-wrack, 204.
Short Commons, 49.
Silurus, European, 56.
Six-spot Burnet, 119, 151.
Snake-stones, 37, 61, 94.
Spiders, 24, 36, 39, 86, 143, 151, 206, 213, 215, 239, 256, 282.
Spiracles of Insects, 199, 254.
Splitters and Lumpers, 73.
Strange Remedies, 85.
The Deep, deep Sea, 169.
Tit in Moustaches, 26.
Toads, 12, 62, 87, 111, 114, 233, 256.
Tom Tidler's Ground, 126.
Variations in British Plants, 32, 228.
Vipers, 2, 95, 108, 131, 143, 160,.191.
What do Crickets eat ? 113, 128, 166,
What Katy did, 146.
What to seek and what to avoid in the Choice of a Microscope, 267.
What's your Hobby? 1.
Why objects appear larger through the Microscope, 8, 45.
Window Gardens, 92, 117, 141, 284.
Wren, the Blue, 199.

Hardwicke's Science-Gossip, Vol. 2. Selection from the Index.

Amœba, the, 223.
Analogy of Form, 266.
Ancient Toads and Frogs, 47, 69, 94, 117, 141.
Anecdotes, a Chapter of, 221.
Ants, 89, 150, 213, 238, 272.
Aquaria, 14, 21, 22, 46, 66, 69, 74, 95, 104, 164, 166, 191, 215, 239, 260, 261, 262,
Bath Bricks, 81.
Beauty, a thing of, 73.
Bees, 17, 22, 47, 70, 71, 115, 119.
Beetles, 41, 71, 88, 89, 183, 190, 279.
Bell-flowers, 219.
Belted Kingfisher, 26.
Bethlehem, Star of, 115, 136, 163, 186.
Bittern, the Little, 200, 277.
Bouquet of Grasses, 53.
Caddis Larvæ, 95, 109.
Canada Balsam, 175.
Captive Owl, 4.
Cats, 63, 88, 255. 260, 276.
Caterpillars, 133, 161, 176. 182, 213.
Chapter of Anecdotes, 221.
Corallines and Acalephs, 124.
Cowslips, 153.
Crown Animalcule, 253.
Death-watch, 34, 75, 254, 278.
Desmidiaceæ, 101, 147.
Diatoms, British, 62, 87, 108, 112, 133, 162, 182, 281.
Electric Fishes, 268.

Fairy-ring Champignon, 225.
Ferns, 46, 61, 95, 164, 173, 269.
Fossil Plants, 37.
Fossil Wood, 250.
Fossil Wood in Flint, 15.
French Marygold, 163.
Galls and Gall Insects, 165, 215, 228.
Gill Fans of Sabellæ, 29.
Glow-worm, 15, 238, 243.
Grey Mullet, the, 145.
Hairs, Star-shaped, 248.
Illuminators for High Powers, 32, 65, 66.
Imperfectly-developed Plants, 8.
Insect Fungi, 127, 176.
Insect Vovarium, 80, 118, 207.
Insects, 41, 55, 91, 93, 112, 117, 147, 161, 184, 185, 204, 213, 249, 282.
Jute; what is it ? 84.
Ladybirds, 169.
Lizard, the Common, 79.
Mistletoe of the Oak, 152, 186, 212.
Moa, the, 14.
Mounting, 19, 20, 23, 33, 47, 93, 94, 114, 125, 209, 245, 260, 263, 282, 283.
Odd Fishes, 50, 99, 171.
Our Club, 193.
Periodic Phenomena, 49.
Pests, and their Checks, 54.
Pin-centres and Rose-centres, 106.
Puff Balls, 270.
Rural Natural History, 83, 163.

Salamander, Japanese, 130.
Saw-fly, the Great, 181, 273.
Scales of Insects, 55, 91, 112.
SCIENCE-GOSSIP, 1, 135, 158.
Serpents at Meals, 244.
Shooting Stars, 274.
Snails and their Houses, 195, 230.
Spectrum Microscope, 52.
Speedwells, 121.
Spider-crabs and their Parasites, 178, 211.
Spiders, 7, 22, 40, 63, 119, 138, 141, 169, 189, 201, 209, 213, 255, 277, 279.
Spring, 97.
Starch, 34.

Sticklebacks, 5, 45, 153, 165.
Teachings of Natural Science, 251.
Track of the Pygmies, 154, 203, 239, 26.
Turtle, the Edible, 247.
Under the Snow, 25.
Vegetable Caterpillars, 176.
Vegetable Fibres, 10.
Venus, Observations on, 110.
Vivarium for Insects, 80, 118, 207.
Viviparous Fish, 241.
Wasps, 22, 88, 116, 276.
Water-Beetles, 183.
Water-Fleas, 156.
Wings of British Butterflies, 27.
Woodpecker, 6, 41, 95, 119.

Hardwicke's Science-Gossip, Vol. 3. Selection from the Index.

A Century ago, 127.
Age of Niagara, 139.
Amidst the Ruins, 31.
Analogy of Smell, 59.
Aquaria, dust on, 69, 117, 118, 141, 142.
Atropos, 51.
Bitten by a Viper, 175, 199, 213.
Blindworm, the, 179, 260.
Blood Beetle, the, 27, 62, 71, 94.
Bouquet from Helvellyn, 242.
Bugs, 269, 276, 282.
Century ago, 127.
Chignon-fungus, the, 107.
Cholera-fungus, the, 206.
Cockroaches, to Kill, 166, 212, 280.
Cornish Colloquies, 182.
Crickets, how to get rid of, 263, 279, 281.
Crocodile in England, 7, 41.
Death's-head Moth, 190, 213, 214, 262.
Death-watch, the, 29.
Dermestes? what is, 28, 206.
Diatoms, 9, 35, 81, 91, 103, 115, 133, 156, 190, 188, 228, 269.
Disguises of Insects, 193, 233, 234, 261, 279.
Dodo, the White, 5, 52.
Dragon-fly, the, 225.
Echoes from the Club, 231.
Edible Bird's Nest, 39.
Fangs of Spiders, 237, 270, 276.
Feast of Roses, 145.
Foraminifera, 36, 129, 215, 236, 263.
Freshwater Sponge, 247.
Freshwater Sticklebacks in Seawater, 38.
Germination of the Toad-rush, 150.
Hairs of Dermestes Larva, 28, 206.
Hardy Foreign Ferns, 83.
Helps to Distribution, 244.
Hints to Object-mounters, 91, 139.
Huxley on the Study of Natural History, 73.
Is Lichen-growth detrimental to Trees? 241.
Leaf-mining Larvæ, 169, 212.
Left by the Tide, 217.

Lichen Dyes, 266.
Maple, Aphis of the, 204.
May Mushrooms, 112, 136.
Melicerta, N. E. Green on, 33.
Mermis nigrescens, 221.
Monmouth Deposit, 133, 156, 180.
Mosquitoes, 78.
Mounting in Balsam and Chloroform, 8, 21, 23.
My little Green Monkey, 179.
Nest of Wood Wasp, 247.
Organization of Mosses, 249.
Oxlip, the, 137, 163, 165, 187, 235.
Perils of a Naturalist, 15.
Poduræ, 45, 53.
Preservation of Fossils, 22, 23, 44.
Primeval Britain, 198.
Primroses, 42, 114, 136, 141, 167, 235.
Ramble in South Africa, 154.
Rare Birds, Shooting, 69, 93, 94, 160.
Rhythm of Flames, 49, 95.
Rural "Folk-lore," 177.
Rural Natural History, 86, 117, 118.
Sandstone, Markings in, 20.
Skeleton Leaves, 22, 117, 141, 246.
Skeleton of Purple Urchin, 82.
Snakes, 273.
Sociable Mites, 124.
Something to do, 99.
Spiders' Nests, 37, 40.
Spirogyra, 60.
Sponge-washings, 228.
Sticklebacks in Seawater, 38, 87.
Swallows, 50, 101.
Temperature of Lakes, 272.
Toad-flaxes, 201.
Travellers' Tales, 25.
Unity of Mankind, 110, 152, 173, 245, 269.
Viper, the, 22, 175, 199, 280.
Wasp, Sting and Poison-gland of, 60.
What's in the Honey? 30, 68.
Wheat Mildew, 16.
White Dodo, the, 5, 52.
Winter Work, 1.

Hardwicke's Science-Gossip, Vol. 4. Selection from the Index.

"Ackersprit," 248.
Agricultural Ant of Texas, 1.
Animals that never Die, 16, 40, 62, 106.
Ants, 23, 59, 88, 117, 118, 138, 143, 159, 177, 190, 213, 234, 261, 263, 282.
Astrantia major, 194.
Birds, 9, 34, 41, 64, 88, 95, 160, 185, 191, 257.
" Black Jack," 232.
Bugs, 17, 31, 46, 214.
Butterflies, Plumules of, 44, 137, 186, 214, 239, 269.
Caddis-worms and their Cases, 152, 189.
Century ago, a, 183.
Cockroaches, 15, 22, 215, 239.
Collecting-bottles, 111.
Cuckoos and Hedge-sparrow, 113, 143, 161, 167, 214, 261.
Cuckoo.spits, 158.
Daddy Longlegs, 256.
Daphnia, the Heart of, 227, 279.
Darwinism ? what is, 241.
Death-watch, the, 87, 113.
Double Eggs, 117, 226.
Dragon-fly, the Pupa of a, 245.
Earthquakes, Phenomena of, 217.
Fairy Rings, 221.
Ferns, 43, 161, 162, 183, 187, 213, 231, 237, 238, 240, 261, 263, 281.
Forget-me-nots, 97.
Formation of Fern Seeds, 183.
Fossil Teeth, the, 53.
Freshwater Actinia, a, 247.
Frogs, 41, 69, 94, 206, 213.
Furze Mites, 49, 114, 160, 209, 271.
Giants, Traces of the, 55.
Gossamer, 51, 58, 124, 143.
Grasses, on the Study of British, 197, 224.
Grasshopper, the Large Green, 196, 236.
Hairs of Indian Rats, 25.
Hawfinch, the, 109, 160.
Hawthorn, Variation in the, 267.
Hedgehogs, 23, 69, 81, 100.
Hobby, the, 229.
Holly-tree, the, 107.

How Birds and Insects fly, 9,
Infusoria, 44, 57, 125, 155, 164.
Irritability and Sensation, 25.
Kingfisher, the Common, 204, 234
King of the Rats, 135.
Kite, the, 251.
Maine Deposits, 85.
Maple Blight, 136, 188.
Merlin, the, 156.
Metamorphoses of Insects, 35.
Microscopic Seeds, 253.
Mole Mite, 232.
Mosquitoes, 207, 211, 212, 215, 236.
Oxlip, the, 35.
Palates of Mollusca, 20, 200.
Pebble-finding, 134.
Perley's Meadow Deposit, 131.
Phantom Larvæ, 78.
Polyzoon from Victoria Docks, 255.
Primroses, Pink, 43, 65, 147, 187.
Reptiles, from the Coal-Measures, 104, 142, 167, 214.
Reptiles in Confinement, 272.
Sand Wasp, the, 205.
Scalariform Tissue, 276.
Sensorial Vision, 55.
Silver-mining in Eastern Nevada, 193.
Slug Parasite, 274.
Smew, the, 55.
Spicules of Echinoderms, 175.
Spiders, 8, 11, 21, 22, 23, 24, 41, 47, 51, 58, 63, 82, 124, 128, 143, 161, 165, 167, 195, 213, 238, 261, 262, 253, 281, 282.
Splits, 169.
Spring Phenomena of Plant Life, 121.
Stag-Beetle, the, 109.
Stings and Poison-glands of Bees, &c., 148, 205.
Storm-glass, 24, 93, 117, 143, 167.
Trees, Age of, 202, 231, 259.
Unity of Mankind, 6, 34.
Vegetable Hairs, 11, 101.
Vipers, 23, 46, 70, 95, 165, 180, 212.
Waxwing, the, 181.
Why ? 265.
Wood-sorrel, the, 52, 289.

Hardwicke's Science-Gossip, Vol. 5. Selection from the Ind

Acherontia atropos, 220, 257, 278.
Admiral, the Red, 257, 262, 278.
Age of Fish, 141, 214.
Animal from the Salt Lake, 78, 130, 234.
Aphis Lion and Lacewing Fly, 15.
Ash-Trees, Violets under, 91, 116, 117, 166, 188, 189.
Badger and Otter, the, 90, 118, 137, 258, 262, 277.
Botanical Allusion in Tennyson, 91, 116, 117, 166, 188, 189.
British Birds, 39, 85, 107, 113, 156, 179, 227, 253, 274.
Buds as Objects for Winter Study, 34.
Butterflies, 58, 116, 140, 164, 212, 273, 274.
Cats and Starfishes, 214, 234, 239, 263, 282.
Cats before a Storm, 117, 141, 164, 167.
Celandine, 52.
Cells for Microscopic Objects, 139, 236, 260, 281.
Centipede, a Luminous, 46, 47, 69, 71.
Cheyleti, 5.
Christmas and the Microscope, 44.
Christmas Berries, 13, 94, 138.
Cilia, about, 53.
Coal, the Story of a Piece of, 1, 46, 71, 96.
Cochlearia officinalis, 43, 66, 91, 114, 143.
Comatula rosacea, 209.
Cuckoo, the, 16, 64, 65, 185, 262.
Death's-Head Moth, 220, 257, 278.
Dendritic Spots on Paper, 22, 46, 71, 80.
Diatomaceæ, 22, 61, 67, 72, 92, 109, 139, 158, 163, 183, 187, 220.
Drawing from the Microscope, 87, 139, 165.
Early Birds, 113, 136, 137.
Enchanter's Nightshade, 62.
English Plant Names, 25.
Epistylis, 83.
Fish Scales, 12, 41, 67, 163, 187, 260, 281.
Floral Giants, 9.
Food for Bullfinch, 215, 237, 238.
Fragillaria Crotunensis, 109, 158, 183.
Frog, 63, 76, 161.
Fumart, the, 22, 45, 68.
Garden Decoration, Wild Flowers for, 169, 265.

Geophilus electricus, &c., 49, 65, 71.
Geranium Robertianum, 133, 191, 2
Geraniums, Carpels of, 211, 235, ! 261
Gnats, 16.
Hawk-Moths, 16, 119, 220, 234, 257,:
Hawthorn, 22, 23, 43, 70, 93, 116, 11
Holly, the, 213, 235, 238, 259, 280, 2
Humble Bee, Winter Home of the, 134, 164, 166.
Hybernation of Bees, 41, 90, 93, 164, 166.
Ianthina, 31, 64.
Influence of Light on Insects, 57, 137, 165, 188, 273.
Insects of the Season, 233, 234.
Kestrel, the, 179, 257.
Ladybird, 232, 239, 267, 283.
Lampreys and Lamperns, 145.
Lapwing, the, 107, 167.
Laurel Berries, &c., 47, 70, 114.
Laurel-Leaves, 20, 45, 68, 71.
Leeches, 76, 93, 143, 161, 165.
Lepisma saccharina, 94, 118, 142, 16
Light Attracting Insects, 57, 77, 165, 188.
Ligurian Bees, 213, 237, 256, 263, 28
Luminous Centipede, 46, 47, 69, 71.
Monsters of the Deep, 55.
Mosquitoes, 17, 54.
Moth, the Fish, 94, 118, 142.
Myriapods, 49.
Norfolk, Rare Birds in, 160, 161, 192.
Otter and Badger, the, 90, 137, 258.
Otters, 90, 118, 137, 161, 165, 184.
Phronima, 73.
Poppy Seeds, 11.
Rudd, Scales of the, 12.
Scales of Butterflies, 212, 214.
Scurvy Grass, 43, 66, 67, 91, 114, 14
Sea Anemones, 56, 90, 198, 210.
Sea Birds, Association for the Pro tion of English, 10, 42.
Sections of Fossil Wood, 18.
Siskin, the, 39.
Starfish, Cats and, 214, 234, 239, 282.
Surirella, a New, 61.
Vine Diseases, new, 59.
Violets under Ash-Trees, 91, 116, 166, 188, 189.

Hardwicke's Science-Gossip, Vol. 6. Selection from the Index.

Abstinence of Insects, 64.
Animated Oats, 190, 211, 212, 237.
Antique, Studies from the, 14, 45, 71.
Ants, 87, 185, 241, 263.
Apterous Insects, 49, 105, 126, 164.
Aquarium Difficulty, 40, 121, 142, 165, 179, 191, 215, 216.
Bees, 21, 24, 34, 42, 47, 65, 70, 117, 119, 141, 142, 143, 161, 164, 165, 166, 167, 188, 190, 212, 213, 215, 236, 237, 257, 269.
Beetles, 112, 113, 143, 233, 263.
Birds, 13, 35, 54, 58, 65, 86, 108, 113, 132, 157, 159, 179, 184, 246, 276.
Borago, 165, 189, 212, 213, 214, 239.
Butterflies, 22, 143, 251, 283.
Caterpillar, 47, 70, 149, 239.
Cats, 16, 17, 23.
Charadrius pluvialis, 58, 95, 114, 188.
Chenopodium Bonus Henricus, 189, 214, 234, 238.
Chrysopa perla, 231, 237, 239.
Cleaning Shells, 190, 214.
Cowper and the Nightingale, 142, 175, 209.
Cowslips, 141, 164, 166, 190.
Cuckoo, the, 108, 138, 281.
Derivation of Foxglove, 43, 67, 69, 91, 93, 115, 118, 135, 166.
Diatoms, 22, 37, 55, 61, 140, 178, 235.
Dodder, the, 117, 118, 120.
D'Orbigny's Foraminifera, 81, 106, 155.
Dragon-fly in Town, 237, 262.
Dun Cow, Ribs of the, 23, 38, 63, 69, 94, 96.
Eggs, 92, 114, 132, 185, 191, 212, 251, 259, 283.
Embryos, Tricotyledonous, 115, 139, 145, 226.
English Plant-Names, 127, 210, 211, 227, 234.
Female Moths, Attractive Power of, 153, 174, 213.
Fish, Scales of, 92, 140, 187, 235, 279, 280.
Flies, 65, 249, 273.
Foraminifera, 9, 68, 81, 106, 155, 167, 209.
Foxglove, 43, 67, 69, 91, 93, 115, 118, 135, 166, 260.
Frog, the, 71, 119, 191, 233.
Good King Harry, 189, 214, 234, 238.
Hairs, 111, 116, 212, 235.
Hawk-moths, 114, 161, 209, 229, 232, 233, 283.

"Hoddy-doddy," 21, 70, 93, 94, 212.
Holly Berries in July, &c., 186, 210, 213, 234.
Insects, Brain of, 262.
Insects, 8, 41, 49, 64, 76, 92, 105, 118, 126, 133, 137, 164, 224, 236, 247, 255, 262.
July, Holly Berries in, 186, 210, 213, 234.
Lotus, its History and Traditions, 124, 208, 231, 272.
Manipulation, 20, 187.
Mergus albellus, 95, 110, 137.
Mosquitoes, 167, 168.
Mosses, 83, 139, 275.
Otters, 21, 54, 90, 113, 114, 137, 138, 209.
Phacelia tenacetifolia, 167, 189, 211.
Plant Names, 127, 210, 211, 227, 234, 262.
Plover, the Golden, 58, 95, 114, 188.
Podder, 91, 117, 118, 120.
Polyxenus lagurus, 187, 209.
Pygidium of Lacewing Fly, 231, 237, 239.
Queen of Spain Fritillary, the, 233, 258.
Ribs of "the Dun Cow," 23, 38, 63, 69, 94, 96.
Rue and Rosemary, 39, 40, 67, 118.
"Sembling" of Moths, 153, 174, 213.
Shells, New British, 130, 138, 161, 277.
Smew, the, 95, 110, 137.
Snow Bunting, the, 71, 90, 114.
Sphinx atropos, 114, 161.
Studies from the Antique, 14, 45, 71.
Study of Natural History, 112.
Sundew, Hairs of the, 111, 212.
Titmouse, the Bearded, 65, 95.
Toad-flax, the Ivyleaved, 43, 71.
Toads devouring Bees, 117, 141, 142, 161, 164, 166.
Tricotyledonous Embryos, 115, 139, 145, 226.
Veronica Buxbaumii, 43, 91, 186.
Wasps and Bees—Do Toads eat them? 117, 141, 142, 161, 164, 166.
Wiltshire, Spur-winged Goose in, 51, 95.
Wingless Insects, 49, 105, 126, 164.
Woodpecker, Greater Spotted, 184.
Woodruff, the Sweet, 47, 69, 70, 93.
Zinc Troughs for Bees, 142, 167.
Zonites glaber, 161, 277.

Hardwicke's Science-Gossip, Vol. 7. Selection from the Index

Anecdotes, twice-told, 192, 212, 237.
Ants, 1, 17, 90, 127, 245, 270, 273.
Arge Galathea, Parasites on, 233, 258, 262.
Bat, the Vampire, 233, 277.
Bats, 41, 66.
Bees, 15, 71, 116.
Birds, 10, 63, 92, 150, 159, 161, 209, 214, 226, 236.
Blind-worm, the, 160.
Borax and Cockroaches, 117, 142, 166, 168, 214.
Borrago, 139, 214, 239.
Botanical Exchange Club, London, 96, 114, 117.
Boulder, the Story of a, 5, 94.
Budget of Queries, 143, 166, 167.
Bullfinches, Captive, 154, 183.
Buxbaum's Speedwell, 114, 139.
Cammocke, 114, 142.
Canine Predilection for Fruit, 263, 283.
Caterpillar, 65, 95, 193.
Cerastium, Abnormal, 259, 279.
Chicken with four legs, 258.
Cleaning Skeletons, 165, 191, 213, 239, 262,
Cluster-Cup, New British, 156, 188.
Cockroaches and Borax, 117, 142, 166, 214.
Cockroaches, 168, 190, 212, 168, 214.
Cotsaold Lion, 119, 142.
Crimson-speckled Footman, 234, 239, 262, 277.
Deiopeia pulchella at Brighton, &c., 234, 239, 262, 277.
Diatomaceæ, British, 69, 105, 188.
Dogs and Fruit, 263, 283.
Earthworms, 118, 142, 143, 166, 167, 189, 212, 262.
Earwigs, 94, 116, 119.
Echinodermata, Pedicellariæ of, 42, 119.
Eel-pout, 20, 30.
Eggar Moth, the Small, 90, 113, 116, 165, 213, 237, 238, 257, 262, 263.
Eggs, 21, 32, 70, 93, 94, 118, 207, 237, 238, 262, 263, 283.
Elephant Parasite, 131, 185, 211, 234.
Eyestones, 21, 46, 69, 89, 93, 95.
Fish in the Jordan, 166, 189, 213.
Fish Scales, 20, 44, 140, 164, 188, 236, 260, 280.
Fleas! Fleas! 97, 155.
Floras, Local, 163, 187, 212, 214, 259, 279.
Flowers and Insects, 258, 282.
Folk-Lore, 213.

Fox-Moth Larvæ, 263, 276, 283.
Fungi, 23, 43, 69, 77, 91, 118, 188, 236.
Gardeners, Early, 19, 43.
Gentian, 91, 119, 139, 143.
Golden Oriole, the, 158, 258.
Gorgoniadæ, 52, 92, 112.
Hairs of Plants, 83, 204.
Hawfinch, 137, 184, 212, 213, 239, 262.
Heartsease, 43, 163. 165, 191.
Hemp Agrimony, 116, 189.
Idolocoris elephantis, 131, 185, 211, 234
Is the Landrail a Bird of Passage? 45, 70, 71, 90, 94.
Kestrel's Egg, 237, 262.
Larvæ of Fox-moth, 263, 276, 277, 283
Lepidoptera, Eggs of, 70, 93, 94.
Longevity of the Goose, 112, 167.
Lotus, the 19, 118, 142, 166.
Luminous Plants, 121, 191, 243.
Lung of the Frog, 92, 120.
Marine Aquaria, 196, 256, 281.
Missel-thrush v. Squirrel, 131, 189, 21, 237, 238, 256, 257, 278.
Moths, Processionary, 106, 185, 188 225, 209, 239, 263, 276, 277, 283.
My Kestrel, 62, 88.
Oolitic Fossil Plants, 157, 212.
Orange-tip Butterfly, 161, 208, 238.
Ornithological Queries, 119.
Otters, 17, 89, 161, 189.
Oxlip, Primrose, 115, 133, 163.
Pineapple, the, 82, 114, 117, 143, 187.
Posting Natural History Specimen 191, 215, 235, 259.
Processionary Moths, 106, 185, 20 239.
Protective Mimicry, 204, 248.
Skeletons, Cleaning, 165, 191, 213, 23 262.
Small Eggar, 90, 113, 116, 165, 213, 23 257, 263.
Soles, Bait for, 237, 261.
Sounding Apparatus, 117, 137.
Spiders, 12, 18, 35, 46, 152, 215, 231.
Star-fish, Pedicellariæ of, 42, 119.
Sunshine, Bat in, 161, 215.
Tamarisk manna, 45, 70.
Tortoise Eggs, 208, 263.
Trichiurus lepturus, 17, 88, 113.
Tritons, 142, 166.
Turbid Aquarium, 93.
Water-snake, 142, 165, 167.
White Ants, 1, 90.
White Varieties, 191, 261, 210, 235, 23 281.
Wryneck, the, 87.

ence-Gossip, Vol. 8. Selection from the Index.

40, 46, 184, 238, 70, 93.
208, 216.
114, 261, 263, 264.
165.
47, 62.
, 212, 238, 239,

13, 79, 114, 116,
74, 184, 187, 214,

1, 31, 65, 89, 95,
9, 236, 241, 280.
ne, 234, 239, 249,

95, 115, 140, 163,

4, 115, 137, 142,
0, 215, 237, 262.
, 240.
1, 69, 144, 183,

, 39.
14, 166, 190, 214,

17.
ring, 25, 49, 73,
3, 199, 217, 241,

118, 162, 175, 212.
175, 237, 261.
3, 232, 239.

3, 116, 140, 163,

llows, 57, 143.
, 71, 72, 256, 276.
, 40, 46, 184, 238.
nicide? 70, 116.
71, 91, 92, 137.
14, 143, 164.
1, 118, 281.
, 9, 137, 190, 283.
e of the, 108, 142.
s, 71, 94, 119.

een, 59, 92, 119,

142, 232, 259.
; Sand, 69, 144,

162, 187.
114.
f Spain, 139, 237,

13, 116, 164, 185,
7.

Gas-light: does it kill Plants? 118, 141, 142, 191, 212.
Gipsy-moth, 23, 69, 119.
Glass-rope Sponge, 35, 56, 106, 181.
Glow-worms, 68, 91.
Golden Eagle, 115, 165, 258.
Gold-fish, 93, 139, 165.
Gryllus viridissimus, 59, 92, 119, 141, 182, 238.
Hawfinch, Gossip, 130, 189.
Heracleum giganteum, 215, 237.
How to stock a pond, 119, 142.
Hydras, Fresh-water, 132, 237.
Ichthyosaurus, 43, 92.
Insects, 12, 18, 21, 87, 89, 139, 161, 185, 212, 234, 236, 256, 257, 278, 282.
Larvæ, 161, 199, 215, 234, 283.
Lepidoptera, Colours of, 175, 237, 261.
Lesser Pettychap, 93, 118, 143, 167.
"Liver," the, 22, 40, 41, 69, 93, 163, 212.
Lunatics and the Moon, 116, 163, 190.
Mice and Birds, 116, 214, 231.
Microscopical Difficulties, 164, 165.
Montagu's Harrier, 167, 189.
Mosses about London, 11, 35, 64.
Moths, 9, 22, 23, 45, 47, 69, 89, 115, 119 121, 137, 139, 163, 164, 166, 190, 214, 230, 236, 238, 241, 280.
Naturalists' Clubs, 46, 64, 115, 185.
Naturalists, Work for, 18.
New Books, 34, 83, 129, 200, 242.
Nightingale, the Irish, 198.
Noctuas, Sugaring for, 115, 141, 164.
Oak Eggar Moth, 22, 47, 137.
Optical Phenomenon, 213, 263.
Origin of the Game-cock, 164, 236.
Pacific Deep-sea Explorations, 186, 211, 260.
Parasites on Starling, 213, 237.
Passion-flower, the, 95, 113, 117, 166.
Phosphorescence, 142, 257, 262, 282.
Pigeons, Wood, 93, 116, 140, 165, 236.
Podura Scales, 16, 41, 47, 100, 209.
Poisoning dried Plants, 40, 70, 90.
Preserving Fungi, 116, 185, 193, 210, 239.
Resurrection Plant the, 213, 231.
Rotatoria, new Species of, 9, 112, 256.
Rotifer, Parasitic, 112, 137.
Rural Natural History, 215, 237, 253, 282.
Saffron, 21, 107, 141, 143, 163, 164, 166, 199, 215.
Scientific Guide-books, 142, 188.
Sections of Vegetable Substances, 177, 231, 255, 282, 283.
Shamrock, the True, 113, 138, 142.
Showers of Frogs, 20, 143, 167.
Silver, Arborescent, 17, 47, 62.

Hardwicke's Science-Gossip, Vol. 8—continued.

Singing Mice. 47, 65, 94.
Snake, the Smooth, 208, 232, 239, 258.
Spicules of Sponges, &c., 20, 95, 188.
Spiders, 140, 182, 213, 261.
Sponges, 17, 20, 35, 56, 95, 106, 160, 185, 281.
Stag Beetle, 42, 92, 95, 139, 190.
Starlings, 93, 117, 118, 140, 213, 237.
Swallows, 57, 89, 115, 143, 212, 235, 237, 239.
Tissues, on Staining, 111, 120, 136.

Test-scales of Podurœ, 16, 41, 47, 10 208.
Tufted Duck, the, 70, 94, 143, 164.
Unicorn, the, 46, 68, 95.
Vitrina pellucida, 44, 69, 116.
Wasps, 70, 163, 214, 239, 282.
Web-weaving Caterpillars, 58, 94, 11 142.
Whitebait, 44, 263, 281, 282.
Young Caterpillars in Confinement, 16 188, 237, 262.

HARDWICKE'S SCIENCE-GOSSIP,

*Published Monthly, price 4d.   Each Volume can be had separatel
Price 5s., in purple cloth. All the numbers are kept in print.*

Volume 9 commenced January, 1873.

MR. HARDWICKE begs to inform Authors Works on Natural History, Travel, Gener Science, and Miscellaneous Literature, that he h at his command the requisite means for bringi1 all Works published by him prominently before t] Public, both at home and abroad. Being practical acquainted with Printing, and having been mar years engaged in business requiring an intima knowledge of the best modes of Illustration, he enabled to offer great facilities to Gentlemen w] entrust their Works to him.

Estimates of Cost, Terms of Publishing, and oth particulars on application.

LONDON:
192, Piccadilly, W.

WYMAN AND SONS, PRINTERS, GREAT QUEEN STREET, LONDON, W.C!

www.ingramcontent.com/pod-product-compliance
Lightning Source LLC
Chambersburg PA
CBHW030356230426
43664CB00007BB/609